New Perspectives on the South

Charles P. Roland, *General Editor*

This Land, This South

An Environmental History

Revised Edition

ALBERT E. COWDREY

THE UNIVERSITY PRESS OF KENTUCKY

Copyright © 1996 by The University Press of Kentucky

Scholarly publisher for the Commonwealth,
serving Bellarmine College, Berea College, Centre
College of Kentucky, Eastern Kentucky University,
The Filson Club, Georgetown College, Kentucky
Historical Society, Kentucky State University,
Morehead State University, Murray State University,
Northern Kentucky University, Transylvania University,
University of Kentucky, University of Louisville,
and Western Kentucky University.

Editorial and Sales Offices: The University Press of Kentucky
663 South Limestone Street, Lexington, Kentucky 40508-4008

Library of Congress Cataloging-in-Publication Data

Cowdrey, Albert E.
   This land, this South : an environmental history / Albert E.
Cowdrey. — Rev. ed.
      p.   cm. — (New perspectives on the South)
   Includes bibliographical references and index.
   ISBN 0–8131–1933–2 (cloth : alk. paper). — ISBN 0–8131–0851–9
(pbk. : alk. paper)
    1. Human ecology—Southern States—History. 2. Southern States—
Description and travel. 3. Southern States—History. I. Title.
II. Series.
GF504.S68C68 1995
333.73′0975—dc20   95-36831

This book is printed on acid-free recycled paper meeting
the requirements of the American National Standard
for Permanence of Paper for Printed Library Materials.

"The landscape . . . either developed or wild,
is an historical document."

RODERICK NASH

Contents

Editor's Preface	ix
Acknowledgments	xi
Note to the Revised Edition	xiii
Introduction	1
1 Isolation and Upheaval	11
2 The Problem of Survival	25
3 The Uses of the Wild	45
4 The Row-Crop Empire	65
5 Exploitation Limited	83
6 Exploitation Unlimited	103
7 Conserve and Develop	127
8 The Transformation Begins	149
9 South into Sunbelt	169
10 Myth and Dream: An Epilogue	197
Notes	201
Bibliographical Note	225
Index	233

Editor's Preface

"History, not geography, made the solid South," once wrote a distinguished southern scholar. Yet no explanation of the region's uniqueness can afford to ignore the effects of its climate and soil. An early Virginia historian accounted for the indolence of his fellow colonists by saying they were "climate struck." "Where God almighty is so merciful as to work for people, they never work for themselves." Ulrich B. Phillips opened his famous book *Life and Labor in the Old South* with the sentence, "Let us begin by discussing the weather, for that has been the chief agency in making the South distinctive."

The story Mr. Cowdrey tells brilliantly here is that of the human response to the southern environment and that environment's response to the humans who settled in the region. He presents the environment as both friend and foe. His story ranges from deeds of ruthless exploitation to acts of dedicated preservation.

In writing this book, the author draws upon years of studying the subject. Because the story provides much of the matrix for the region's larger history, it is most fitting for a volume in "New Perspectives on the South." The series is designed to give a fresh and comprehensive view of the South's history, as seen in the light of the striking developments since World War II in the affairs of the region. Each volume is expected to be a complete essay representing both a synthesis of the best scholarship on the topic and an interpretive analysis derived from the author's own reflections.

CHARLES P. ROLAND

Acknowledgments

Pepys apart, very few people have written a book unaided, and it is one of the pleasures of authorship to acknowledge one's debts. The subject of this essay was suggested to me by Charles P. Roland, who deserves credit for launching the endeavor, though no blame for the faults of the final product. Since the story obliged me to intrude upon a variety of specialized scholarly turfs, I was fortunate in finding Wilcomb Washburn ready to read my remarks upon Indians, and Wayne D. Rasmussen, my first attempts at agricultural history. John Duffy's comments on the medical aspects of the subject were pertinent and welcome. Lawrence B. Lee read the entire manuscript with an eye both sympathetic and critical. Oliver A. Houck—formerly general counsel of the National Wildlife Federation, now of the Tulane Law School faculty—granted me a lengthy interview, read the draft of the final chapter, and criticized it from the viewpoint of a most knowledgeable environmentalist. Joe Hughes and Carl L. Tyer of the Weyerhaeuser Company were patient and informative while leading me through some of their employer's woodlands on a memorably hot day. John Spinks of the Endangered Species Office of the Department of the Interior provided me with some essential data. Gaines M. Foster, now of Louisiana State University, read the early chapters and provided incisive comments. Alice Crampton, a doctoral candidate at American University, aided me as a diligent and responsible research assistant; she also typed the manuscript. Without her help I might have completed the book, but not, I think, during the present century.

I suppose the problems that face a historian—the perennial amateur, the jack of all trades who is master of none—when he attempts such a study as this one will be evident to all. The election of 1980 clearly marked a major change, if not a counterrevolution, in the nation's environmental policies. In its aftermath, people seem to abound who combine daunting intellectual mastery of their particular specialty with religiose commitment to one or another resource strategy. In such an atmosphere the best thing, perhaps, is to present the essay which follows as an exercise in the somewhat neglected virtues of humility and the long view. Southern history is a good place to learn both, and some hope, as well.

Note to the Revised Edition

The words with which I concluded the foregoing section seem to me as apt in 1994 as they were in 1983. In other respects, the events of the intervening years have required me to make some changes in *This Land, This South*.

Increasing consciousness of pollution and the burden of pollution control and increasing though still insufficient efforts to save the wetlands have marked the years since the first edition. (The ongoing debate over the reality, causes, and long-term significance of global warming is a regional concern only insofar as it is everybody's concern.) While some major construction projects with substantial environmental impact have gone to completion, new ones seem increasingly unlikely. A friendlier attitude toward the environment in the executive branch has replaced the bitter hostility of the days when the book first appeared, though performance has by no means kept pace with the promises of either George Bush or Bill Clinton, and the impact of a conservative Congress remains to be assessed. The hot-eyed passions of the early 1970s, when my own interest in the environment formed, seem extraordinarily remote, with the movement's concerns now institutionalized, not only in the Environmental Protection Agency but even in the Army Corps of Engineers, which many environmentalists then considered to be The Enemy.

If I were beginning the book now instead of emending it, I might write it differently; it seems to me I might have integrated a discussion of pollution-caused ills into my account of the region's natural diseases and shown one curve of incidence rising as the other sank, both changes ironically reflecting different sides of the same scientific revolution that has remade the South as an incident to its main job of remaking the world. I might have written a chapter on urban ecology, perhaps along the lines of William Cronon's work on Chicago; the importance of the city as a crucial environmental fact in an urbanizing age occurred to me as I was writing this book, but I went to work on other things and never pursued the subject. That remains for others to do.

This Land, This South has taken its basic form, and I am no more likely to try to change that than I am to attempt amateur surgery on an old friend. The reader of this new edition will find an expanded bibliographical note and a rewritten chapter 9, as well as appropriate changes to the footnotes and the index, designed to bring the story roughly a decade forward. I have tried to make the book more relevant without changing its original aim, which was to survey the environmental history of a complex region through a long history in a little book and, hopefully, to issue a *vade mecum* to others who may pass beyond it.

This Land, This South

Introduction

TO THE SPANISH SAILORS WHO, IN THE YEARS BEFORE 1500, FOLLOWED THE north-flowing currents of the Florida Strait toward home, the land which they glimpsed (and considered part of Asia) must have seemed a mere line, a darkening of the horizon mists. Then as now, the sky over the green Gulf took a Himalayan form when squalls threatened. The rookeries of the Keys—gulls, herons, and terns—were in fuller voice then, and the seamen may have seen some animals, such as the West Indian seal, which have since become extinct or nearly so. Some seamen paused on the Keys to kill turtles and gather eggs, then sailed on.

Inland, and soon to be discovered by these harbingers of change, stretched a region which differed from other parts of the continent in both physical and cultural forms. On time scales appropriate to either, it possessed a remarkable age. "Columbus," wrote J.H. Parry, "did not discover a new world; he established contact between two worlds, both already old."[1] The South of rocks and weather, which abides, and that of men, which has changed utterly, both confirm this judgment.

The skeleton of the eastern highlands rose through a series of orogenies, beginning about 475 million years ago. Such mountain-building episodes are believed to result from the collision of continental masses, in this case between North America (or rather its primitive form, Laurentia) and a number of fragments created in the splitting by tectonic forces of an earlier megacontinent. Sequential collisions with these "small . . . fragments or island arcs" may explain the complex structure of the mountains, in which metamorphic rocks deformed by heat and pressure overlie sedimentary material which once formed the eastern margin of the continent.[2]

Varied forces have since worked on these materials. None have been more basic in their effects than rain and heat. The South lies mostly sunward of the isotherm that marks a 60° Fahrenheit mean annual temperature, though the highlands, the border states, and Virginia are significant exceptions. Rainfall is in excess of forty inches a year, though most of Texas and part of the border states of Oklahoma and Missouri receive less. Rain and heat are both makers and destroyers of soil. Heat favors the rapid decay of organic matter; rain produces erosion in all its forms and the leaching of minerals from the earth.

Forest cover also helped to form the soils. In recent times much of the South has been covered by a great forest—pines in a broad belt paralleling the coast, a wider growth of hardwoods in the Piedmont, mixed woodlands in the swamps and river bottoms, and a succession of forest zones in the mountains, where altitude mimics increasing latitude. Beneath its trees the region is largely covered by forest soils called podzols—brown in the bluegrass regions and Great Valley, which are rich in lime; red and yellow in most other areas. Along the coast are water podzols and marsh soils sometimes enriched by decaying beds of ancient marine shells. In the river valleys are dark, fertile, unconsolidated soils brought by floods, as well as swamps floored with tenacious buckshot clay. In mid-Alabama, and stretching into Mississippi, is a region of Tertiary soil underlying a fertile grassy region, and a similar formation belts eastern Texas. But the red and yellow podzols predominate, stretching from within sight of the Potomac almost to Houston, from the Missouri almost to the north shore of Lake Pontchartrain, and from southern Kentucky to eastern Florida.[3]

The chief peculiarity of southern soils, except the alluvial, is that they are old. Indeed, the South might be defined as the part of eastern, humid America that escaped the last glaciation. Soils south of the Ohio River lack the topping of minerals carried and ground to powder by the ice and distributed (along with unwelcome larger stones) over the Northeast. Except on the eastern rim of the Mississippi valley, the South also lacks the secondary gift of the glaciers—the deep loess, wind-carried from the glacial region, that forms the basis of the midwestern prairies.

Exposed to leaching under heavy rains for ages longer than northern topsoils, the southern earth has progressed farther in a chemical evolution that tends over time to reduce soils to a mixture of clay and the hydroxides of aluminum and iron. Organic decomposition, fed by the leaves and needles of the forest, forms carbonic acid that carries soluble nutrients deep beyond reach of the roots of many plants. By speeding up chemical reactions, heat causes all processes to move more rapidly. Where valuable minerals such as limestone exist, the processes of enrichment also proceed faster; but in most of the South geology has given little that can enrich. Organic materials tend to be vaporized by the heat, and less of their nutrient content enters the soil than in cooler regions. Nor does the ground freeze for long periods in most of the South, trapping nutrients above the ice.⁴

Caution must be used in discussing fertility, a notion almost as debatable as beauty. By comparison with the Utah salt flats and similar desolations, the pine-barren is a cornucopia. Complex natural ecologies can thrive in spots that agriculture pronounces hopeless, for nature, lacking special purposes, does not demand wheat where only thistles grow. Fertility in the most restricted sense of yielding good crops to human effort depends not only on mineral wealth but on soil structure and tilth, on the state of agricultural science, on the diligence of the farmer, and not least on his expectations of what the land should yield. Yet over time the South's natural dower of soil, of which so much is "mediocre in quality, highly erosive under intense cultivation," can only have imposed a constant, tenacious drag upon the growth and wealth of human societies so long as they drew the basis of their livelihood from the land.⁵

An essential part of the South, yet geologically distinct, is the Mississippi valley. Here land is new rather than old—indeed, the river is in process of creating the landscape. When Pleistocene glaciers shut off the drainage of the midcontinent through the St. Lawrence valley, rivers were diverted into the Mississippi embayment. The melting of the glaciers brought vast new quantities of alluvium, much of it washed from the debris of the ice sheet. From its head at a gap in the Commerce Hills of Missouri, the alluvial valley took form as a great wedge of earth and sand, its blunt end turned toward the Gulf. Over this the river took a variable course, building natural levees by its overflows, eroding its banks to form meander loops, cutting off the loops in time of flood to create oxbow lakes. Beneath the valley, faulted bedrock made for instability and occasional earthquakes; the most severe ever recorded in North America occurred near New Madrid, Missouri, in 1811–1812.

A part of the coastal plain in many ways, the alluvial valley shares with it a restlessness that is rare in real estate. A rising sea level caused by the melting of the icecaps erodes both the Gulf and Atlantic seaboards; the sea level has risen in some areas as much as six inches since the early twentieth century. On the Atlantic coast the embayed section from Cape Fear to the Chesapeake, plus the Sea Island area, is sinking, while the so-called Cape Fear Arch, for obscure reasons, may be on the rise. On the Gulf coast erosion devours marshlands, while the Mississippi delta threatens to change its location.

Helping to change the land, especially the coast, is southern weather, which favors the melodramatic. Storms are a yearly fact of life. Many summer days feature the pregnant pause and the wild onslaught: the palmetto fan pauses while the prostrate fanner gapes at hailstones bounding in the street. Tornadoes are commonest in the Mississippi valley and the Gulf coastal plain. "During the thirty-year period ending in 1950," notes a physiographer, "every fifty-mile square of this area was struck by about a dozen tornadoes, and in the whole of the area more than 2,000 persons were killed." Hurricanes are makers and unmakers of the coasts. Hardest hit are the barrier islands, whose outlines are redrawn decade by decade, changing shape, as Lafcadio Hearn wrote of those near the mouth of the Mississippi, "more slowly, yet not less fantastically than the clouds of heaven." Studies in the Florida Everglades have revealed how greatly hurricanes can change the land. Storms in 1935 and 1960 killed extensive mangrove forests and washed soil and vegetation from some beaches while leaving massive deposits on others; bird rookeries were devastated and tens of thousands of birds killed. When the debris had dried, fire swept it, killing most of the living mahogany. On the Gulf plain such epics of destruction have been written some 150 times since the Europeans first came; the Atlantic has been less often but no less brutally visited.[6]

The climate is, however, kind to life in many forms, including some that most humans would rather do without. The vivid fauna of the marshes and the still vigorous wildlife of the southern woods are not the whole story. Heat and damp encourage a legion of buzzing, creeping, leaping, and gnawing insects. The intimacy of southern people with a meager biting bug, the mosquito *Anopheles quadrimaculatus,* and with the parasites of malaria, brought in the post-Jamestown era a toll of tedious suffering and economic loss that is beyond computation. By comparison, the eruptions of yellow fever with its vector *Aedes aegypti* have been mere episodes, though memorable ones, sufficient to give many areas, for a century or two, a recurring economic malaise which was a disease in itself. Pathogenic organisms often are favored by heat because they are not so likely to die in the passage from one host to another; their vectors are commoner and their lives in the vector easier. Malaria is highly responsive to temperature, its range lying roughly south of the 60°F summer isotherm. Carried into the North and West during the colonial period, malaria by the Twentieth century had with few exceptions spontaneously retreated to the southern valleys and lowlands. Similarly, epidemic yellow fever after about 1830 was exclusively a southern disease.[7] Yaws, hookworm, dengue, and filariasis have afflicted the South. African and European parasites colonized the region with their hosts, and, like them, sometimes became makers of history.

As in the case of southern soils, a note of caution must be sounded in any discussion of the southern disease environment. The extreme upper and mountain South experiences winters of considerable severity, and sharp though brief periods of hard frost can strike anywhere but extreme south Florida. As natives like to point out, there is no barrier between south Texas and the North Pole except one fence, and the fence blew down. Wintry episodes help to control insect life even as they bring health problems of their own. The general if spotty mildness of the winters, on the other hand, puts considerably reduced stress on all life, benefitting both animals and humans, especially the old. Yet when all is said and done, the South has been gravely burdened by disease, and since Jamestown, usually burdened beyond the American norm. An old, warm, and wet region largely covered with mediocre soils and a host to abounding life, the elemental South, since long before the first discovery by European man, has modified the health, wealth, and character of the societies that have occupied it.

After the lapse of 500 sometimes catastrophic years, it is hard to say which is more remarkable, that so much has changed in the southern landscape, or that so much remains the same. To fly over the present-day South is to be struck by the fact that after so long an encounter with civilization much of it remains, or has become again, a great forest scattered with habitations. The mountains abide except where much coal has been found. The Mississippi River has been disciplined and usually keeps between its levees, but only provided they are watched all the time and seasonably shored up with money. Despite some centuries of market hunting, feather hats, DDT, and heavy-metal effluents, most of the wild species survive somehow. Some only eke out existence under closest protection; others, like the deer, increase in the wild, culled by regulated slaughter. Meanwhile, invaders other than man—primordial or obnoxious creatures, most of them, such as the armadillo and the fire-ant—expand their domains, exploiting man's tyranny or ignoring it altogether. Some insects are positively mocking in their unkillableness, in their ability to become more unkillable as the fields are heaped with toxins.

Even today nature does not exist solely to support man's extravaganza. It remains willful, mysterious. After centuries of effort, men are still trying to figure out that highly stochastic process, the Mississippi River. Sometimes they guess what it will do next, sometimes not. The ancient belief that trees, beasts, and rivers are gods has at least this to be said for it, that they tend to follow their own sweet will, for all of man's knowledge and power. Yet man, who made the dachshund out of something very like a wolf, has taken an imperious way with the southern landscape. Any environmental history must find an inevitable theme in the increasing human power that tends toward both a more refined stewardship and a more dangerous inadvertency. Intriguing is the extent to which the natural environment of the South, including much that is usually termed primeval, is an artifact of sorts, shaped if not invented during the millenia of human occupation. Corn was formed in the Indian's garden, the forest transformed by his burning. The deer were ambiguously his cattle or his prey. The processes by which the land was reshaped for human purposes are of long standing, and in some ways the impact of the earliest interventions remains greatest.

It does appear that one artifice demands another, as theorists have said, the process building until the artificer grows weary and longs for a little spontaneous nature. Yet he is constrained by his own handiwork. A marked irony of southern history has been the extent to which the home of states' rights has needed and pursued federal subsidies, and the locus of a durable anti-intellectualism has come to lean upon the crutch of science. To manage its poor soils the South needs the best agricultural know-how; to conquer its formidable roster of diseases, advanced medicine and sophisticated health-delivery systems; to make its famous climate more asset than curse, all of the above. When a few more billions must be appropriated for the levees, when the fields groan with pesticides, it is easy to dream of an Eden of spontaneous nature. But in fact man the changer has been so long at work in the South that even the pre-Columbian forest bore his mark.

Nature and history make odd conjunctions. No doubt the latter is a special case of the former. Yet the two modes that divide man's uneasy being live by different time scales and are shaped by different kinds of evolution. Biological processes are Darwinian, rooted in the universal facts of differential reproduction and survival, which mold with infinite slowness the plastic species. Human society's evolution is Lamarckian, working by the transmission of acquired characteristics.[8] Struck by war or revolution, society may suddenly revert to the Darwinian mode, all the trappings of culture stripped off by the needs of survival. When culture is strong, human purposes revise nature, upending the logic of survival so that simplicity is favored over diversity, the helpless domesticate over the hardy wild thing. Such changes have been a part of the South's history, which is long enough so that neither biological nor social revolution is anything new.

I hope that the reader of this essay will find in it some hint of the bond which southerners have formed with their spot of earth. Besides being a subculture, a web of traditions, or a sequence of events, the historical South is a place, the life that is lived there linked beyond separation to a certain ridge, a certain river, a certain quality of days in summer. Its artists have affirmed what all southerners, including the numerous separated brethren, know in their bones—there is no life apart from place, and no place is exactly like any other anywhere. The link between man and land is not all definable, and perhaps what matters most is least recorded. The intense, almost physical bond that ties the southerner to his place seems to develop independently of whether that place is grand or seemingly beneath notice. There is scenery in the South—the Smoky Mountains, for instance, are awash in it. But throughout the coastal plain and Piedmont a monotony of fold and hollow, of wood, marsh, and meadow predominates. Yet from wiry grass and pine thickets, from a sun-struck spiderweb or a misty day in autumn, what a wealth of declamation has been drawn! The South, like other regions long inhabited, has become a landscape of the mind.

Perhaps a word is needed on the scope of this essay, geographical and otherwise. I early decided to attempt a survey of the interface between culture and nature in the region, for if "environmental history" means anything it seems to mean that. Certain stories had to be told that are customary in such accounts—what happened to forests and wildlife over time, how industrialization affected pollution, and so forth. The subject also seemed to demand treatment of a topic usually consigned to the history of medicine, namely the incidence of disease and some of its effects upon the human resources of the society it afflicted. I took as my point of departure Philip D. Curtin's useful phrase "the disease environment." I could not see any aspect of man's encounter with natural forces of much greater importance than his long travail with microbes. The links between soil, diet and the diseases of malnutrition also seemed to make the latter a part of the story.

The breadth of the subject was somewhat daunting. I resorted to selectivity, using the essayist's excuse that his form demands it. The reader will consequently find some dozens of pages on the development of the Mississippi River, nothing on the Pee Dee, and a mention or two of the Pearl. I could have written a bigger book and included more, but I do not think I could have excluded the topics treated here.

Southernness was a problem. So was weather. Ulrich B. Phillips opened his history of life and labor in the Old South with the words, "Let us begin by discussing the weather, for that has been the chief agency in making the South distinctive." The Florida anthropologist William H. Sears recently wrote that "the present states of Georgia, Alabama, Florida, Mississippi, Louisiana, and Tennessee form a neat unit geographically, ecologically, and culturally, today and in the geologically recent past." Actually I have not found the unit to be quite so neat, nor even the weather so consistent or instructive. What is one to do with the Canadian forests of the higher altitudes in the Smokies, or with the drier central and western parts of Texas? Again, since this is the story of the encounter between a particular subculture and a particular environment, what about the parts of the South that are only ambiguously southern—the farmers of western Maryland and the Great Valley, whose agriculture, like their ancestry, is Pennsylvanian; or the state of Maryland itself, which was a part of the South once but is no longer? Both culture and nature usually lack hard edges; they fade away cinematically, as the successive southern cultures have faded since Indian times into the sunset over the great plains. I decided in the end to adopt a similarly twilit method, to let my attention settle upon Maryland for a time and then withdraw, and not to attempt the prairie at all. Dryness makes for another agriculture, if not another world; here I had to draw a line.[9]

Because the evidence used in this study is both abundant and spotty, the reader may find too many facts in one place, too much speculation in another. The landscape of the essay, like that of its subject, is a patchwork of well-worn fields and fallow. Subject to my own limitations, I have tried to show some of the ways that man and land have shaped each other in a little corner of the world.

1

Isolation and Upheaval

THE HISTORY OF THE WESTERN HEMISPHERE IS ONE OF SEPARATION FROM THE Old World broken by intermittent rediscovery. Because there are penalties for prolonged isolation, the two great discoveries of which we have knowledge entailed ecological revolution.

The origin of the American Indian was long a mystery and in many respects remains one. He is a cousin of the Mongoloid peoples, the descendant of a common ancestor. His original home was in northeastern Siberia. He sprang, it would appear, from homogeneous stock, perhaps a few closely-related tribes of big-game hunters who preyed on the great fauna of the Pleistocene. (The Eskimo is a Mongoloid from a later incursion.) Indians are believed to have entered the vast cul-de-sac of the American continents during the glaciations when sea levels were low and a land-bridge connected Alaska with Siberia. Some certainly came during the period from 28,000 to 10,000 years ago, and others may have arrived much earlier. They spread rapidly over a truly pristine continent where no human had been before them. Some may have entered the South. Crude scrapers and choppers found in Alabama, as well as archeological sites in South Carolina and Texas, suggest, though they do not prove, very early human occupation—on the order of 38,000 to 40,000 years ago.[1]

From about 11,000 B.C. to about 6,500 B.C., America was populated by Paleo-Indians, people at approximately the cultural level of the Upper Paleolithic in the Old World. Fluted (grooved) spear-points are their major monument. So too may be the catastrophic Pleistocene extinctions, for their arrival and progress coincided with the destruction of dozens of species of large animals. When the Folsom hunters of 8,000 B.C. ranged the Llano Estacado of western Texas and New Mexico, these lands were green savannahs, with wooded valleys hiding streams and marshy spots.

In this pleasant landscape the animals were ambushed, killed, and butchered. Dwelling on ridges that overlooked the watering-holes, the hunters' world was a movable feast as they followed the herds, warring against elephants, horses, camels, sloths, and bison. The Indians used ambushes and natural traps such as ravines, driving their prey with fire. Lightly armed with stone-tipped thrusting spears, they may by such methods have killed many animals to secure the meat of a few.[2]

Whether they actually destroyed the great fauna remains in doubt. The die-off also correlates with the drying of the plains. Yet the notion of climatic change overtaking swift-moving animals like the horse and camel, well able to survive in arid regions, strains belief. Evidence is persuasive that man did not evolve in the New World; he erupted suddenly as a new force, armed with weapons of stone and fire. His prey had no opportunity to adapt gradually to the great killer. In this view man's coming exemplified, in Charles Darwin's words, "what havoc the introduction of any new beast of prey must cause in a country, before the instincts of the indigenous inhabitants have become adapted to the stranger's craft or power." The Pleistocene extinctions, however, can only have been a most complex process. Climatic change and human hunting acted simultaneously but in very different ways to transform ecological balances. Man may have administered the *coup-de-grace* to species under pressure from an altered habitat. By destroying carnivores, he may have set off population explosions in some species of herbivores which then crowded out others by competing for diminished food sources. The possibilities are many, the evidence spotty. Early man participated in the extinctions, but we do not know how.[3]

In any case, the consequences were a major calamity for the later complex Indian societies. Their domesticates—dog, guinea pig, llama, guanaco, and vicuña—were a poor lot by comparison with the domesticates of Eurasia. Animal muscle is a prime mover of civilization. Cultures devoid of the horse and bullock, however brilliant their expedients were at a permanent disadvantage in peace or war against others so equipped. The horse, domesticated from Asian stock, reentered the North American continent bearing a conquistador on its back; its role in the conquest was a considerable one. Some of its descendants, running wild in a hospitable environment, became the mounts of the plains Indians and an essential ally in their hunting, warring, empire building, and resistance to the whites—in short, in the renascence of their culture. What Montezuma's generals might have done with light cavalry, to say nothing of war elephants, remains a teasing dream.

Paleo-Indian hunters may have entered eastern North America during the Two Creeks interstadial, about 10,000 B.C., when the drying of the high plains forced the big game eastward. (At this time "the East" meant the Southeast plus the Ohio valley and the Atlantic coast to Maine.

Glaciers still covered the rest.) Here the Indians found and attacked the tree-browsing mastodon. The East is littered with elegant Clovis spear points that mark their coming. Early sites occur in present-day Texas, Alabama, Florida, Missouri, Tennessee, North Carolina, Kentucky, and Virginia.[4]

By the eighth millennium B.C., the extinction of the great fauna—the mammoth, horse, and camel had all vanished in the ninth millennium, *Bison antiquus* a thousand years later—began to bring into existence the durable cultural forms which archeologists call the Archaic. Indians turned to a mix of small-game hunting, fishing, and wild-plant collecting. In some places wandering hunters attached themselves to unmoving food sources, such as mussel beds, and turned to a semisedentary lifestyle. The marks of an aging, presumably more populous culture were plain in the intensive exploitation of more limited areas, and in the diversified food sources—deer, raccoon, opossum, birds, fish, shellfish, wild vegetables.[5] This tradition was not only a temporal phase but an enduring aspect of Indian life; in remote and poor regions the Archaic lingered past the time of Columbus.

Meanwhile, a third Indian tradition, the Woodland, took shape along the Mississippi and Ohio rivers after 1,000 B.C. Agriculture, including the first use of corn in the region, cord-marked or fabric-marked pottery, and elaborate burial customs characterized the Woodland world. Immense earthen mounds were used for burials, as at Poverty Point, Louisiana, or to form enclosures, or to represent the shapes of animals. Notable centers of culture flourished near Chillicothe, Ohio, and in the temple mounds at Kolomoki in Georgia. Woodland Indians cultivated native plants as well as the squash and a tropical variety of corn which had diffused northward from Mexico. But gathering was still an important activity—wild nuts, fruits, berries, and seeds like those of the ragweed were important to their diet—and they remained hunters, harrying deer and smaller game.[6]

The Mississippian tradition rose along the middle reaches of the river in the floodplain between Vicksburg and St. Louis. During its 800-year reign (ca. 700–1500 A.D.) it spread to river valleys throughout much of the Southeast. The highest culture north of Mexico in its day, the Mississippian was characterized by a firmer agricultural base—the growing of corn, beans, and squash, in particular—by its riverine locations, its large towns defended by palisades studded with wooden towers and sometimes by flooded moats. Within the walls, mounds topped by temples or the homes of chiefs dominated a stratified society.[7] Elegant tools and carvings and a well-developed religion were among the Mississippian achievements.

The most formidable impact on man and environment, however, and the most enduring, lay in establishing corn (*Zea mays*) as the basis of agriculture. Maize probably derived from a wild plant now called *teosinte* in Meso-America 8,000–15,000 years ago. The plant was almost certainly transformed by human selection, "man's most remarkable plant-breeding achievement," in the words of one authority. The modern plant is a biological monstrosity, unable to survive unless human hands separate the kernels from the cob and plant them. (Otherwise the seedlings are so dense where the ear falls that competition may prevent any from reaching adulthood.) The inability of corn to scatter its seeds is a mark, as striking in its way as temples or pyramids, of Indian artifice in the transformation of nature.[8]

The "most efficient of all cereals in converting nutrients and solar energy into foodstuff," corn made possible the great Indian civilizations, the provincial but impressive cultures north of Mexico, and the relatively dense Indian population of some 96–100 million that may have existed in the Western Hemisphere before the coming of the Europeans. The whites in turn would find it "the most useful grain in the World; and had it not been for the Fruitfulness of this Species, it would have proved very difficult to have settled some of the Plantations in America."[9] As a food source, *Zea mays* has one notable failing in that it fixes an important vitamin, niacin, in a chemical bond that resists digestion. Bean plants intercropped with corn provided an important supplement, assisting the return of nitrogen to the soil and supplying a necessary addition to the human diet, as well. Mississippian sites were confined to river bottoms, probably in consequence of the need to grow corn without fertilizer, whose use the southeastern Indians, unlike those of the Northeast, never learned. In the lowlands they fished and hunted, traded, and spread the forms and insights of their culture by boat, enduring floods for the sake of the fresh alluvium brought by the waters.[10]

In the later Indian South, the three traditions coexisted, enabling different communities to survive in a variety of natural settings. At the same time, the unities of Indian culture pervaded the region. Their use of fire to clear land for planting and to drive animals during the fall and winter hunting transformed the landscape. Fire became central to the maintenance of a human-centered ecology. Burning discouraged larger forest fires by clearing deadfall and brush; it provided nutrients for trees; it kept the woodlands open, encouraging growth of grass, and making acorns easier for animals, birds, and men to find. It helped to create and preserve meadows whose dense, shrubby margins were the hiding place and browsing place of the deer. Culling the deer herds by hunting provided a sound control on the species, whose caution, speed, adaptability, and prolific breeding protected it from extinction. Indeed, the deer by breeding too fast would soon have starved themselves without predation both human and nonhuman. Deer were the bison of the southeastern Indians, and in some measure their cattle, both prey and symbiont.

Fire helped shape the face of the land in ways that struck European travelers and explorers. Even before they reached the coast, they might smell the smoke of burning woods, as did one off Delaware in 1630, who wrote that "the land was smelt before it was seen." Inland a few miles from the Virginia shore they found heavily timbered areas interspersed with savannahs of "many hundred acres"covered by tall grass and spotted with bogs where trees still grew thick with interlacing branches. In many places the land was a deer park like the one spied by Barbadian explorers along North Carolina's Cape Fear River in 1663: "We found a very large and good Tract of Land on the N.W. Side of the River, thin of Timber, except here and there a very great Oak, and full of Grass, commonly as high as a Man's Middle, and in many Places to his Shoulders, where we saw many Deer and Turkies; one Deer having very large horns and great Body, therefore we called it Stag-Park." Much of the upland region was open, with meadows of "fine-bladed grass six foot high, along the Banks of . . . pleasant Rivulets," where the naturalist John Lawson saw "buffalos and Elks" at the turn of the eighteenth century.[11]

The woodlands of longleaf pine may be the work of man. In the natural succession, the longleaf forest gives way in time to a mixed growth of gum, oak, loblolly, and short-leaf pine. Where the longleafs maintain themselves as the dominant tree, plant scientists adjudge them a "fire climax"—the product of an incomplete succession interrupted by fire. (Because of its thick bark, rapidly growing root system, and seeds which germinate on exposed soil, the longleaf is unusually though not perfectly fire resistant. Hence its competitive advantage in woods that are frequently burned.) The supposition that millennia of Indian burning halted the forest succession over large areas is rational, if in the nature of the case unprovable.

The habit of burning, like that of eating corn, has proved durable. In colonial times (and later) a number of laws were passed which forbade the practice. But southerners gave little heed to the law; on the evidence of the repeated enactments, burning continued and has not ended today. A 1943 study concluded that "in the vegetation of the Southeast, effects of fire are more striking than in any other region of the world." Fires, of course, can have destructive effects; they may promote erosion by destroying ground cover or kill the longleaf pines themselves if set too often. A study in the 1960s found "by far the greatest loss" of American forests to fire was in the Southeast, a third of it attributable "to fires purposely set as part of a long-established practice (whether good or bad) to help establish new crops of pines, to improve grazing, and to help reduce the fire hazard."[12] Fire, like agriculture, to which it was closely allied, reshaped much of the South in line with the needs of its human masters.

The less tangible aspects of the Indian past are not without interest for the historian of the South. To read studies which treat broadly of the whole Indian culture is to be left, among so much that is strange, with a sense of eerie familiarity. The southeastern Indians, it would appear, not only did much hunting and ate a corn-centered diet. They also drank great quantities of a blackish fluid rich in caffein—though as an emetic—, loved violent games, doted on oratory, were great devisers and purveyors of myths, and had a taste for blood vengeance and a legal system (or nonsystem) based on private justice. Influence of a common land? Parallel development of two peoples? Enough said. Such matters are beyond footnoting. But the eerie feeling does arise.

The value of Indian ways cannot be seen only in what has survived; much of what was best was quick to perish. The general satisfactoriness of Indian life to those who lived it must have been great despite its hardships and the cruelty of its wars. The advantages of their highly adapted lifestyle are suggested by estimates that they were able to live off the country much of the time, providing by manual labor only about one-fourth of the year's subsistence. Recent studies have confirmed that a mixed regime of hunting and shifting agriculture forms the most energy-efficient of all economic systems, requiring only .05 to .10 calorie of input to produce a calorie of food energy. This efficiency, of course, depends on a large ratio of land to population, as well as on the gearing of effort to subsistence. As noted by one Virginia historian in the eighteenth century and another in the nineteenth, the effect of the maladapted culture introduced by the Europeans was "to make the native pleasures more scarce."[13]

The varied life and limited labor surely contributed to the Indians' extreme reluctance to be converted to white ways. Early settlers repeatedly commented on the great difficulty of persuading Indians under any condition to adopt Christianity and civilization. In 1728 William Byrd II remarked on the "bad success" of the charitable in conveting Indians, who, "after they returned home, instead of civilizing and converting the rest . . . have immediately lapsed into infidelity and barbarism themselves." English racism was not the whole reason. The Jesuit father Pierre de Charlevoix, an astute cleric who traveled through Canada and Louisiana, roundly declared that "there never was so much as a single Indian that could be brought to relish our way of living," a fact which he attributed to the liberty and naturalness of their lives.[14] Perhaps their initial resistance was as much a tribute to Indian good sense as the later success of some tribes, especially in the South, in adopting (and adapting) white ways when they had become convinced that no other way to survival lay open.

These reflections should not, however, be taken to endorse a myth of our own time, the notion of the Indian as natural ecologist. Writers nowadays like to speak of the Indians' "balanced ecological systems," but the historian may wish to enter a partial dissent. The successive appearance of Paleo-Indian, Archaic, Woodland, and Mississippian traditions suggests not only that the southeastern Indians were evolving complex cultures, but that they were not part of a balanced, self-maintaining system. Like the whites and blacks of a later day, they pressed against the land's resources and reshaped its forms in line with their own desires.

Yet by the time the Europeans came, Indian ways had long attained a high degree of subtlety and sophistication. Methods of hunting, farming, and gathering had been fine-tuned by millenia of learning, and complex attitudes had been embodied in myth and parable for transmission to the young. One myth taught that disease was caused by animals in vengeance for human cruelty, a mark of ecological conscience that had the merit of conferring power on the powerless, always a strong stimulus to caution in the mighty. Whether the living world of the Indian called forth an ethic in any sense comprehensible to modern man is quite another matter. Indians appear to have dealt directly with their environment, in which animals and natural forces were selves, exerted power, and demanded respect. They did not have the luxury of regarding the natural world as a thing.[15]

The Indian society that the Europeans found was old, complex, and subtly adapted, and its members, though neither nobles nor savages, knew how to perpetuate its modes of intimacy with nature through taboo and the poetic precision of myth. They were also engineers, people of artifice who had helped to kill off the great beasts, and had spread the pine forest, domesticated corn, transformed the character of the land by fire, and spread it with the mounds and carvings that bemused their successors. In some exceedingly fundamental ways they had shaped the matrix of later cultures. Their most durable achievements suggest goals which the white and black society may hope to attain also, given time.

Testimony on the Indian world in its afternoon came from the white explorers who began to penetrate the South in the sixteenth century. Though by 1546 the Mississippian culture had already developed signs of decline, it still impressed Hernando de Soto with its population, abundance of corn, and capacity for war.[16] Instead of lingering in the comfortable corn-rich villages, however, his band made a fighting traverse of the inland South and ultimately found the Mississippi: "The river was almost halfe a league broad. If a man stood on the other side, it could not be discerned whether he were a man or no. The river was of a great depth, and of a strong current; the water alwaies muddie; there came down the river continually many trees and timber, which the force of the water and streame brought downe. There was great store of fish in it of sundrie sorts."[17] Beyond, among the Tulas of Arkansas, the Spanish reached the borders of the southeastern culture area. Here the Indians lived mainly on buffalo and raised little corn, an echo of the big-game hunters and their culture, which in the Southeast had passed away or been extensively transformed some ninety centuries before.

Cabeza de Vaca's tale of his wanderings and survival between 1528 and 1536 provides a view, at a man's eye-level, that sweeps from Florida to Texas. In Florida he found Indians living a peasantlike existence among maize fields in a landscape of great trees and open woods. Here too were wild beasts both expected and unexpected. ("Among them," he wrote, "we saw an animal with a pocket in its belly.") Birds were everywhere. Bison came from the north as far as the coast, and the Indians hunted and ate them, used their hides for blankets, shoes, and buckles, and traded hides to the interior. But the Texas coast, where his adventures ultimately brought him, was another world, where the Indians had neither "maize, acorns, nor nuts," went naked, planted nothing, and were hungry much of the year, supporting life on the staple of the American deserts, the prickly-pear. They controlled their numbers by avoiding intercourse for two years after a birth and suckling children to the age of twelve. Here was a remnant of Archaic life, and a poor sort of it at that, on the fringe of the more developed Indian world, and a sharp reminder of the complexity of that world.[18]

The conquistadors passed through the Indian South like bullets through cotton, leaving no trace but their chronicles, and seemingly no mark of lasting consequence save their animals. Visitors more tenacious and ultimately more destructive began to explore the Atlantic coast later in the century. To a surprising extent the efforts of the English were guided by a rational determination to understand and exploit the southern climate and the crops which the soil might be expected to produce. Their efforts on Roanoke Island from 1585 on have been termed by a leading scholar "an important scientific experiment to work out what kind of settlement it might be possible to insert into a particular ecological setting."[19] They sought a region that would grow what the homeland did not, and, deluded by considerations of latitude, Richard Hakluyt the Elder wrote of Virginia as "a countrey or province in climate & soil of Italie, Spaine, or the Islands whence we receive our Wines & Oiles," where settlers might set about "the making of Wines and Oiles able to serve England."[20]

Methodically the English began learning something of the land and its people. Savants and artists accompanied and indeed led the soldiers, sailors, and adventurers who made the first essays at Roanoke. Thomas Harriot and John White surveyed the North Carolina mainland and the offshore islands, and mapped the coast from Cape Lookout to Cape Henry. They examined Indian horticulture and customs, and learned at least one local language. White drew pictures of the plants, animals, and human inhabitants that helped to fix the image of the New World for generations of Englishmen.[21] Nothing they learned deterred their hopes for "southern" crops, in the European sense, and their conclusions helped prepare the way for major disappointments later on.

A second category of products which the English expected from the New World—ironically, in view of what followed their settlement—was its medicines. Tobacco had won its earliest European fame as a therapeutic. Taking the weed by mouth was supplemented by smoking, a practice which passed from the Indians to seamen and thence to the more adventurous of all classes. During his stay on Roanoke, Thomas Harriot "vsed to sucke" smoke after the Indian manner, and back at home found the practice to be common among "men and women of great calling . . . and some learned Physicians also." He heartily recommended tobacco's curative powers, lending his voice to the promotional activities of his patron, Sir Walter Raleigh. Harriot attributed the fact that the Indians "know not many greevous diseases wherewithall wee in England are oftentimes afflicted" to their smoking.[22]

Other western plants were becoming only less famous than the imperial weed. Guaiacum had been introduced into Europe in 1517 as a cure for syphilis. In the 1530s and 1540s sarsaparilla gained fame just as ill-deserved for the same reason; though Andreas Vesalius treated it with suspicion, Gabriel Fallopius endorsed it, and the public agreed with the distinguished discoverer of the function of the mammalian oviduct. By 1610 the root would be worth £ 200 a ton in London. Sassafras was another miracle plant, hailed in 1571 by Dr. Nicolas Menardes of Seville in his quickly translated *Joyfull Newes out of the Newe Founde Worlde* as a cure for fever. In 1602 an English ship would bring from the Carolina coast, as a valuable cargo, sassafras, smilax roots believed to be sarsaparilla, "Benzoin," "cassia ligneia," and a certain fragrant bark—the last three of unknown nature. Among the lists compiled by the early English adventurers, three classes of vegetable products—the "southern," the forest products and naval stores, and the medicinal, including tobacco—predominated. Together with dreams of gold and silver, and solider hopes for iron and copper, these were the proposed economic bases of the colonies to come.[23]

Led by men of ability and buoyed by what seemed rational expectations, the adventurers found at first that all went swimmingly. An English soldier reported that the Indians were like people of the Golden Age, full of generosity, free of deceit. Early reports breathed the wonder of the new-found land: "Vnder the banke or hill whereon we stoode, we behelde the vallyes replenished with goodly Cedar trees, and hauing discharged our harquebuz-shot, suche a flocke of Cranes (the most part white) arose vnder vs, as if an armie of men had showted all together." A scene on the Carolina coast in the piping summer of 1584.[24]

In 1585 Ralph Lane found North Carolina "very well peopled and towned, albeit sauagely, and the climate so wholsome, that wee had not one sicke since we touched the land here." The last phrase was in the nature of a qualification, for the English had brought sick with them,

Harriot reporting that the expedition of 1585 had only four deaths among its 108 people, and that "all foure, especially three, were feeble, weake, and sickly persons before ever they came thither." These four unfortunates had a role to play in the rapid decline of the Golden Age that followed.

If the great fauna of the Pleistocene were unready for the Indians, the Indians in turn proved tragically unready for the whites. The reason did not lie in numbers, culture, or fighting qualities. Recent estimates suggest that ten to twelve million Indians may have lived north of the Rio Grande, something between one and a half and two million in the Southeast.[25] Although fragmented by tribe and language, the southeastern tribes were clever and warlike, even if their own fights were usually rather a matter of raiding for blood vengeance or captives than Eurasian-style war. De Soto and his men had ample experience of Indian fighting qualities during their Long March through the hinterland. The Indians' lack of metal and large domestic animals represented a serious weakness, though later events showed that European artifacts, if not their methods of manufacture, were easy to acquire in trade. The most serious danger to the Indians from the new contacts was physical, and a direct consequence of their long isolation from the remainder of mankind.

Pre-Columbian Indians were not, of course, free from disease. Skeletal remains of Mississippian settlements show clear evidence of disease, including possible signs of syphilis.[26] Bone lesions rarely yield unequivocal diagnoses, but American Indians suffered identifiably from arthritis of the hip and spine, osteosarcoma, various deficiency diseases, osteomyelitis, bladderstones, and endoparasites whose ova have been found in mummies. From many of the common diseases of Eurasia they were, however, free. Anthropologists see their presumed origin as hunting tribes in a very cold region as the cause of this relative—and dangerous—good health. Northeastern Siberia and the Bering Strait region in Pleistocene times formed a "cold screen" through which comparatively few diseases or their vectors could pass. The New World was without human beings, and, once the land bridge had disappeared, remained unknown to all but its inhabitants, save for random and temporary contacts. Even that pervasive human affliction, malaria, probably did not exist in pre-Columbian America; at any rate, such later pest-holes as the Gulf coastal plain of Mexico supported large populations. The Indians were by no means free of the ills that flesh is heir to, but in regard to many deadly contagions they faced the Europeans as immunological virgins.[27]

By contrast, the Europeans, once substantially isolated themselves, had developed extensive contacts with Asia and Africa, and had paid a considerable price in the Black Death, a gift of the Eurasian caravan routes.[28] They suffered also from vivax malaria and smallpox. In addition, from their earliest visits to the New World, possibly from the voyages of Columbus onward, Europeans carried with them free or enslaved Africans. Natives of a continent singularly rich in diseases, blacks brought a variety of ailments on their involuntary journeys to America.

From this array of ills, the Indians, save for randomly occurring natural immunes, carried no defense in their bodies. The fragmentation of their society, in general so fatal to their efforts to resist the whites, provided a measure of protection. The institution of tribal warfare tended to keep Indian communities separate, the villages and hunting ranges divided by belts of woodland. Since trade, ceremonies, and war itself provided contacts, this could only have been a partial defense, but it made for a pattern of brushfire epidemics, roughly preceding the whites westward from the coast. Over the course of the conquest, however, mortality would be appalling. Seventy percent of the Indians whose tribes had contact with the whites in the seventeenth century may have died. "Disease," wrote a recent historian, "literally destroyed much of the native population of America and altered and shaped the customs of the survivors." Compared to this elemental force, the bullets and pikes of the Europeans were of small consequence in bringing about the conquest of America.[29]

The first epidemics probably occurred in the northeastern regions, where European fishermen worked the Newfoundland banks and dried their fish ashore. Disease followed soon enough in the South, as well. Here exploration by the Spanish began in the 1520s, St. Augustine was founded in 1565, and exploratory colonies of French Huguenots and Englishmen were briefly planted during the same century. Outbreaks of illness occurred among Indians in the vicinity of the Spanish settlements in Florida. In 1526, Lucas Vasquez de Ayllon established a short-lived colony in the vicinity of the Cape Fear River, which was abandoned within six months because of hunger and disease among the settlers. In 1540 Hernando De Soto encountered the "lady of Cofitachequi," an Indian queen or dignitary who received him into a comfortable, prosperous region rich in corn on the Savannah River. Here his chronicler, the Gentleman of Elvas, noted: "This countrie was verie pleasant, fat, and hath goodly meadows by the rivers. Their woods are thin, and full of walnut trees and mulberrie trees." But "within a league, and halfe a league about this town, were great towns dispeopled, and overgrown with grasse; which shewed that they had been long without inhabitants. The Indians said, that two yeeres before there was a plague in that countrie, and that they removed to other townes."[30]

Whether the settlement of Ayllon had established disease in the region is unclear. The coastal tribes were, however, in danger from multiplying contacts. In 1585 Sir Francis Drake raided St. Augustine, bringing with his men a contagion—probably typhus—from the Cape Verde Islands.

The disease promptly spread among the Indians: "The wilde people . . . died verie fast and said amongst themselves, it was the Inglisshe God made them die so fast."[31]

Such was the background to the experiences Harriot related of the North Carolina coast. Friction between the English and the Indians there was probably inevitable, for Indian good nature could not have lasted forever in the presence of hungry, acquisitive guests who seemed disinclined to go. There were episodes of violence, quarrels between the tribes complicated by new quarrels with the English, and some shootings. But a fundamental factor in the worsening of relations was the outbreak of unfamiliar and deadly disease under circumstances that inclined the Indians to conclude that they had been "shot at, and stroken [struck] by some men of ours, that by sicknesse had died among them."[32] Meantime the English, viewing with amazement the spread of pestilence that affected only the Indians and not themselves, felt they were in the presence of a miracle. "There was no towne," wrote Harriot,

where wee had any subtle deuise [device] practised against us, . . . but that within a few dayes after our departure from every such Towne, the people began to die very fast, and many in a short space, in some Townes about twentie, in some fourtie, and in one sixe score, which in trueth was very many in respect of their numbers. . . . The disease also was so strange, that they neither knewe what it was, nor how to cure it, the like by report of the oldest men in the Countrey never happened before, time out of minde. . . . This marueilous [marvelous] accident in all the Countrey wrought so strange opinions of us, that some people could not tell whether to thinke vs gods or men, and the rather because that all the space of their sicknes, there was no man of ours knowen to die, or that was specially sicke. . . .

Struck by the prodigy, some Indians prophesied that more of the strangers would come to kill them and take their places. Those who were to come were in the air, invisible, and without bodies, and it was these who made "the people to die in that sort [pestilence] as they did, by shooting invisible bullets into them."[33] The Indians tried to starve out the English, and sometimes killed them. The English used their weapons freely, but above all they retaliated more harshly than they knew merely by being there. The desertion of the first English colony was played out against this ominous background, as was the fate of the Lost Colony after 1587.

For the South the sixteenth century was one of foreshadowings rather than accomplished changes of great dimension. Nevertheless, a "transforming experience which was in the end to be total" had begun, and the "human adaptation to and compromise with the natural environment which had been under way for at least 20,000 years" was ending.[34]

The early ships carried more across the Atlantic than men and women and disease. The Spanish had set the pattern, for they embarked, like Noah, carrying their animals to stock the new land. Horses and cattle, gone wild, made their own unrecorded marches of exploration north through Mexico. Entering Texas in the eighteenth century, Spanish settlers would find longhorn cattle there before them. On the ships came also seeds and plantings—wheat, rice, woad for blue dye, sugar cane, radishes, lettuce, figs, citrus, and vines. Weeds, grasses, and ornamentals were carried hopefully by settlers or rode as stowaways; in time, Kentucky bluegrass, daisies, and dandelions—immigrants all—would colonize the American fields.

Some of the imports, of course, failed to take. But many reacted to the new environment with explosive growth. By 1555 Englishmen could read in the translation of Gonzalo Fernandez Oviedo y Valdez's *Historia General y Natural de las Indias* how "all such sedes, settes, or plantes" brought out of Spain and planted in Hispaniola became "muche better, bigger, and of greater increase" than in Europe. Similarly, the beasts expanded in the manner of exotics suddenly turned into a region where natural enemies were few, and grew "much bygger, fatter, and also of better taste then owres in Spayne." The spreading cattle industry would soon claim more hands, and possibly produce more wealth, than the famous mines of Meso-America; both Spanish and English would bring it to the South, and kine would partially replace deer in the savannahs and open woodland, as in a later time they would replace the buffalo on the high plains. In turn, the conjunction of wheat, maize, potatoes, and cattle laid the foundation for the American abundance in the production of food which, over the centuries, was probably to be a greater lure of immigrants than gold or freedom.[35]

No such splendid vistas beckoned the Indian inhabitants of the South, or of America at large. Though for a time the tribal wars, the ceremonies, the making of pottery and lives went on, and the temples on the mounds were presumably full on feast days, the tragedy rehearsed along the coast was not far off. Within a generation there were glades in the coastal South where the fires of autumn and winter were set by new hands. As before, the flames sent wraiths among the gum trees, the red oaks, and the pines, spreading the odor that so mysteriously evokes lost time.

2

The Problem of Survival

EVEN AS THE ENGLISH SETTLERS SEEDED THE NEW WORLD WITH THEIR DISEASES, they began to transform the land and forests by their labor. They faced their new existence in a state of overwhelming weakness—physical, psychological, and economic. The voyage deprived them of strength, the strangeness of everything beset them like an enemy, and the land was recalcitrant about yielding the necessities of life and trade. They suffered terribly, and their experiences did not make them gentle or inclined to hesitate about the misuse of other peoples or the land. Like its virtues, the failings of the society they created emerged under stress and hardened into tradition and law.

The dimensions of their problem appeared early. During 1607 almost half the Jamestown colonists died, and the years immediately following brought worse, not better, mortality. So far as is known, the diseases that killed them—dysentery and (in all likelihood) typhoid—came with the settlers, while another cause, salt poisoning, came from their shallow wells dug beside a tidal river. Yet even when the Virginians had passed through the early years and many had scattered to the interior, illness continued to harry them.[1]

A basic reason lay in the immigrant experience itself. Because people of many backgrounds came together in the ships and in the colonial settlements, all were likely to be exposed to unfamiliar pathogens, against which they had had no opportunity to develop resistance. Many, no doubt, brought on board as well the consequences of a lifetime of inadequate food. Most vulnerable of all were the young, less experienced in disease and probably even less disciplined than their elders in matters of personal cleanliness.[2] In the ships the luckless travelers were shut for months on end to eat tainted and inadequate food and to trade their ills and parasites. Worse were convict ships, where typhus and smallpox killed some 15 percent of the prisoners before they reached America. Among Monmouth's rebels, wrote a witness, "we had enough in the day to behold the miserable sight of botches, pox, others devoured with lice till they were almost at death's dore. In the night fearful cries and groning of sick and destracted persons, which could not rest."[3]

A much vexed special point is the role of malaria, if any, in the Virginians' ordeal. The disease is a masquerader, its symptoms so varied that even today in a malarial region any febrile patient is suspect. By bringing nonimmunes and carriers together in raw towns with nonexistent drainage, colonial settlements favored malaria, and the process of clearing woods and letting in sunlight expanded the areas where the American anophelines preferred to breed. Yet there are reasons to doubt that the disease played any great part in the mortality of the early decades. In seventeenth-century England, vivax malaria—the so-called benign form of the disease—was endemic in certain marshy areas, notably in the counties of Yorkshire, Kent, and Essex. Occasional epidemic outbreaks occurred in London. It seems impossible to associate this usually nonfatal illness with the mortality in early Virginia, and the mortality clearly antedated the first importation of Africans in 1619. Malaria seems, therefore, relegated at most to the status of an adjunct or contributary cause to other and deadlier diseases.[4]

Even after the first hard years, epidemics recurred. In March 1622 the Indians, alarmed by the growing numbers of the whites, fell upon the now-dispersed settlements and killed 347 people. Planting was disrupted and again settlers crowded together in strongpoints. Pollution and concentration had their usual effect; ships continued to bring sickness, disease rates soared, and some 500 died in the year following the massacre. The census of 1624 showed a population of 1,275, as against an estimated 1,000 in 1618, despite the arrival of some 4,000 immigrants during the intervening years. Making allowance for some emigration, a death rate on the order of at least three in four is indicated for this period. Estimates of later mortality include Governor Berkeley's suggestion of four-fifths among imported servants, Robert Evelyn's five-sixths, and historian John Duffy's seven-eighths.[5]

The effects on Virginia were not good, from the viewpoint of its settlers or its promoters. Not only did men die (and women, of whom there was a small minority in the colony by 1618). In hunger they ate up the animals that were to have bred and made for future security. After 1612 many colonists were diverted to Bermuda. Growth of the older colony came slowly. The harsh laws associated with the name of Sir Thomas Dale reflected Virginia's straits. One provision imposed death on any man who killed a domestic animal or fowl, burning on the hand and loss of ears to any accomplice, and "foure and twenty houres whipping" to any who concealed the deed. England's first colony remained a backwater for decades.

The colonists' chief competitors also fell victim to disease. Both before and after the coming of the English, the native pattern of life, in villages of 30–200 persons separated by belts of forest, tended to protect the Indians to a degree. Nevertheless, in 1608, a year after the initial white settlement, a "strange mortalitie" afflicted " a great part of the people" of Accomack on Chesapeake Bay. Indian settlement in the Jamestown area was heavy—according to John Smith, some 5,000, with 1,500 warriors among them, lived within 60 miles. But Smith's own lists of Indian sites suggest a larger population, and early estimates in turn were revised upward as the English became familiar with the region and its many villages. This extensive population was one good reason for the slow early progress of the Europeans, who found the clearings filled by their owners rather than emptied by prior epidemics, as in New England. The dense settlement, however, must have favored the spread of disease.

Sharp fighting occurred between Indians and whites, especially during and after the massacre of 1622. But widespread sickness soon struck the tribes, as well. In 1636 the governor of Virginia wrote that ships were landing filthy and sick passengers—by no means a new complaint—with the result that "where the most pestered shipps vent their passengers they carry almost a general mortality." A direct link between the ships and an Indian epidemic has been alleged in the case of a sailor, infected with smallpox, who landed in what is now Northhampton County, Virginia, in 1667 and passed his disease to local Indians, who "died by the hundred ... practically every tribe fell into the hands of the grim reaper and disappeared, the only exception being the Gingaskins." Meanwhile the colony had admitted its first black slaves. Though their numbers remained small until after 1660, entry of African as well as European diseases may well have taken place. Between 1607 and 1669 the number of the Powhatan Indians declined, from all causes, by about four-fifths. By 1669 the Indian population in the vicinity of the English settlements had dwindled, by one estimate, to about 3,000. In 1728 William Byrd II reported that Virginia's Indians had declined by disease, war, and rum to "a very inconsiderable number" except for the Nottoways, a tribe of 200 members. Some groups had been pushed back through treaties, but the steady emptying of the land was noted also in areas reserved to the tribes. Land acquisition by whites was greatly facilitated as "great tracts ... became deserted without any removal" of the Indians who dwelled there. In 1685 the burgesses noted that tribes living near the Blackwater River had once been very numerous, which was why land had been reserved for them. Now they were extinct.

By about 1650 malaria, very likely in both benign and malignant forms, had become endemic in the Chesapeake. The sapping of strength and increased susceptibility to other diseases which resulted were probably basic to the continuing high mortality. The life expectancy of a colonist of twenty was about nineteen to twenty-four more years; childhood mortality was great, but adult death was so common that many houses sheltered neighborhood orphans. Death gave an extraordinary aspect to the whole society, making it more youthful, throwing open land and opportunity to the young, encouraging them to breed quickly, cutting across the bonds of custom and tradition. A malarial region constantly reinfected with many ills by immigrant and slave ships did not promise long life to any of its people, red, white, or black.

The sword of pestilence cut both ways, but the diseases were incomparably stranger to the Indians, and represented only one (though probably the heaviest) of a battery of stresses afflicting them. The outbreak of epidemic disease in communities where virtually no one had immunity fatally disrupted social ties. Almost everyone was sick at once, care of the sick was impossible, planting was neglected, and hunger seconded disease. Fear of the unknown compounded the disaster. Survivors, especially in a small tribe, easily fell victims to hostile neighbors, or in despair joined friendly peoples, so that their own tride's identity perished.[7]

Meanwhile the numbers of the English were recruited with new vigor from abroad, and after the mid-century the number of Africans likewise rose suddenly. The reason lay in major discoveries that brought the colonists past their time of crisis and in the process gave a distinctive and in some ways tragic shape to their society.

The European settlers came from the sea like lifeforms experimenting with the land but still belonging to the element that tied them to all the things that were familiar. For generations England remained home, and its ocean-spanning empire, young and fragile as it then was, remained their source of supplies and recruits, and the only reason they could hope to make more impression on the vast American continent than the fishermen who had come before them.

The Virginia Company of London was bound to support them, and they in turn must win a cash return for its investors. The natural deduction was that they must find in the new land a product that would be valuable as England understood value. From the beginning this arbitrary demand, unrelated to the nature of the land and even to the needs of the colonists themselves, was imposed upon the new continent. The implication was that the colonists must perceive the land as commodity, not resource. Basic strategies of survival (getting food, keeping warm) imply limited demand: the full belly rejects food, and the warm body declines an extra blanket. But if the new country was a source of commodities to be exploited for sale, then the desires of the sellers would expand with the array of possible purchases, and the struggle to stay alive could modulate with no sense of discontinuity into the pursuit of wealth.

The story of Virginia's survival through tobacco has been often told. From its beginning the colony was a business venture. Its most typical early literary form was the promotional tract, as that of New England was the sermon or spiritual biography. Leaders of the venture broadcast to the sometimes mocking British public the image of a realtor's Eden. Its founders' vision of the New World was of a garden run to seed that wanted only tending to produce bountifully. While the ideal was a rural haven, a broken wilderness transformed into tidy fields, the dynamics of the society, the perils of the early years, and the continuing brevity of life tended rather to encourage brusque exploitation followed by pragmatic enjoyment of comfort once attained. The contrast to early New England was marked, with its religious dynamic, close communities, and rhetorical portrayal of America as a Wilderness of Sinai wherein the Chosen wandered at the Lord's behest. Virginians were at the service of few abstractions; they wanted to survive, then to reproduce in short order the comforts of English rural life, so that those who had been of small estate at home might exist as gentlemen in the new-found world.[8]

The chance of doing so seemed small at first, to men set down among a large and warlike Indian population and harried by disease. Land, the foundation of all, was not to be had easily, at least not in useful form. The Indians had taken the best of it, for they were well experienced in selecting the good soils needed for their corn. Hence John Smith's complaint that "all the Countrey is overgrowne with trees, whose droppings continually turneth the grasse to weeds," referred to the land available to the whites. A brutal job it must have been to clear such land, for people in their physical condition and subject to intermittent Indian attack. So great was the labor, and so extensive and inviting the Indian fields, that some settlers consoled themselves for the massacre of 1622 by reflecting that now the Indians' "cleared grounds in all their villages (which are situate in the fruitfullest places of the land) shall be inhabited by us, whereas heretofore the grubbing of woods was the greatest labour."[9]

Meanwhile the colonists were trying a variety of possible exports. They obtained corn from the Indians as a gift, or by trade or armed robbery. They traded trinkets, utensils, and weapons for deerskins, shipped timber, and continued to trade in medicines, especially sassafras. Attempts to introduce silk came to nought, for though "the wormes prospered excellent well" and the mulberry trees which supplied their food grew naturally in Virginia, seeded by the Indians in "prettie groves," yet "the master workeman fell sicke: during which time, [the worms] were eaten with rats." Experiments continued through much of the century but without establishing a viable industry. Sand was used for glassmaking, and iron, which existed in bogs and in nodules called concretions, was extracted and used in a foundry set up at the falls of the James River. Upland cotton was planted in 1621, the seeds brought from southern Europe where, in turn, the plant had been carried from Mexico. But a perennial problem showed in the fact that the seeds had to be removed from the lint by hand. Flax and hemp were tried repeatedly. Early efforts to grow a tobacco imported from Trinidad failed, as did French grapevines, oranges, and pineapples.[10] John Rolph had better luck with tobacco and began to turn out an acceptable form of the Trinidadian plant, which the export trade demanded in place of the bitter leaf of the Indians. As in the case of skins, medicinals, and fine timber, the colonists had found a product that Europe wanted, and in view of their attempts to establish more useful plants and industries, an ironical success it was.

By the early seventeenth century smoking was well established in the homeland. The market surged, but early Virginia tobacco was an unsatisfactory product. Imported at London in 1614, it was "reshipped by the East India fleets to points where men had learned only to smoke and not to discriminate."[11] The plant must have seemed a poor enough hope, demanding labor that was excessive even by the standard of the times, and this in a land where immigration was slow, the death rate appallingly high, and hands few even to grow necessities. Yet the plant, as many exotics will, thrived in its new home, growing in the layer of mould that overlay the cleared forest soils. The productivity of such land declined rapidly. As the plant demanded labor, it required fertility, of which most Tidewater Virginia soils, other than the alluvial, had little. In a country not required to trade in order to live, a little tobacco might have been grown in the sort of riverine locations favored by the Indians, as a luxury or, as with them, a ritual drug. But for Virginians, production of a salable luxury was life.

Something had to be sacrificed to make tobacco succeed. What Virginia had to excess was unbroken land and the forest that covered it. From very early days the Company had planned to use land as a dividend, spending it as bonuses to bring in settlers. The result was the headright system, usually dated from 1617–1618. Fifty acres of land rewarded anyone who emigrated or brought in another, bond or free. In 1625 the king approved the system by letters patent. By various means, many illegal, individuals could grasp extensive holdings at little or no cost. Engrossment of land began to transform the meaning of the word plantation from a clearing where crops were grown to its later sense of an extensive commercial farm worked by laborers. The conjunction of land acquisition by the headright system with the growing of tobacco by unfree labor became fundamental to the pattern of economic life.

Such farming insured that the minimum number of people would have the maximum impact on soil and forest. Land usage, especially the waste of forests cut down for planting, was excessive because of the voraciousness of the tobacco plant conjoined to the poorness of much of the soil. Beyond the alluvial regions of the Tidewater river valleys, soils were thin and sandy, mediocre and quickly exhausted. Washing away under row-crop cultivation, they exposed "at a depth of three or four inches, a sterile subsoil." The ridges were of clay and sand and "always poor." Tobacco culture was largely limited to bottomlands. Farther inland the river valleys became narrower but the uplands somewhat more fertile, though not uniformly so. Wherever tobacco was grown it required in large to moderate quantities nitrogen, phosphorus, potassium, calcium, magnesium, and sulfur, plus a variety of trace elements. The plant was extremely sensitive to deficiencies; where any one essential element was lacking, the plants were smaller. Nitrogen deficiency resulted in a quick reduction in size above ground, while the root system grew in search of the element; plants lost their green color and the leaves yellowed and dried up. Other deficiencies in major nutrients produced inferior or useless plants, often more susceptible to disease.[12]

It is in these terms that the celebrated "exhaustion" of the soil must be viewed. The nature of the plant and the soil, the expectations of the farmer, and the demands of the market were the functional elements requiring the continual breaking in of new land, which in turn meant the wastage of the forests that, in the planter's view, encumbered it. Good farming in the European sense of carefully husbanding small plots was not to be thought of. The scarcity of labor demanded utmost efficiency in its utilization by confining work to the growing of crops rather than the care of the land. In the off-season, hands were set to clearing new land, when the work did not interfere with the growing of tobacco and food. All trends tended toward dispersal of the population, which the Indians could no longer prevent as their numbers waned under the impact of disease and war. Even transportation—that normally limiting factor—meant far less in the Chesapeake region, so rich in waterways that extended the Atlantic world deep into the landscape.

The planter lived between two millstones, the wants of the market and those of the finicky and demanding weed on which he based his hope of wealth. Care of the plant whose dead and dried end-product must meet the demands of a capricious far-off market ruled his life and those of his family and servants. Clearing, planting, pitching, hoeing, topping, suckering, worming, harvesting, threading, curing, and prizing—every history of the imperial weed runs a roughly similar litany of exact and tedious work required to make a crop.[13] That tobacco drained the soil of nitrogen and potassium was not known to the planter, but that what he considered a good crop could be made for only three or four years on a given piece of land did not escape him. Tobacco land came to mean new land. And only by the continuing sacrifice of land was success possible.

Hence this small farmer (and despite the possession of sometimes extensive acres he usually remained a small farmer, breaking a little of his land at a time to cultivation) kept most of his acreage as a personal soil bank, to be exploited in turn when what was under the hoe wore out.

As the tobacco nexus took form, the planter sought to prevail over many difficulties by other ploys. He resorted to unfree labor, preferring at first white servants with whom he could communicate to heathen blacks. The main exceptions were a few West Indian blacks, neither heathen nor ignorant of western agriculture, who indeed sometimes instructed their masters in the growing of crops. If the planter was lucky and his plantation grew, he became perforce an able taskmaster and a calculating, sometimes brutal manager of the labor of others. Such knowledge as he might have of intensive agriculture he threw overboard as useless in a land where labor must be husbanded and land spent to make a profit. He came to America knowing the use of manure and the plow, but in the root-entangled forest soils he turned to the hoe, and often used no manure at all. (Whether manure did, as some planters claimed, give tobacco a rank taste is unclear.) He turned his stock into the woods to fend for themselves, so that they became half-wild, half-domestic creatures not unlike the deer they replaced. And he viewed with little practical concern such calamities as the winter of 1673 when 50,000 head of cattle were said to have died by cold and starvation. At any rate, he was relieved of the burden of producing food for his beasts.[14]

The planter completed the development of his specialized kind of efficiency by growing corn after the tobacco to feed his servants and slaves (and in part his family and himself). With the forest down and the roots decayed, he was able to use the plow, but the wooden plow, with its six-inch cut, did little to bring nutrients to the surface. Traced straight up and down the rolling landscape, however, its rows made watercourses for the rain. The importance of erosion is not clear in the Tidewater, a country of gentle slopes and absorbent, sandy soils and level bottoms; in all probability its worst effects were delayed until the Piedmont was settled. Wheat proved an especially valuable crop; in virgin soil it failed, growing healthy stalks but little grain, while in the abandoned tobacco fields it could be grown with success. In a word, it filled an available niche. By the end of the seventeenth century, it had become important to Virginia, not only for food but for export.

When all the crops had been made in turn and the land used up, it was abandoned to a long fallow period, optimally about twenty years. This brusque progress of the land, from forest to cropland to broomsedge and little pines within a decade, created the distinctive new look of the Chesapeake Tidewater. As described by a geographer writing of Maryland's South River country, it was "a transient landscape of frequently shifted fields, abundant old fields, and ragged vegetation in various stages of succession."[15]

Yet for the planter who triumphed over the many dangers and obstacles of his life, a more than English existence became possible. He possessed a cash crop of fluctuating value but one which served also as the local medium of exchange, giving him the opportunity, rare among farmers, of growing his own money. Two grains provided basic food, and a variety of vegetables and fruits, native and foreign, supplemented his diet. Despite the neglect, his animals multiplied, as exotics so often do. Though his chance of a long life was small, he and his longer-lived children possessed a genuinely new world. To the lucky and the strong belonged, after the first harsh years, free acres, abundant food, a far more open social order—in sum all the advantages of a less crowded and less used land.

A sign of success was the establishment of Maryland along similar lines, utilizing the Virginians' hard-won experience. From its beginning, the colony projected by George Calvert and his son Cecilius aimed at securing part of the fur trade an establishing a successful colonial venture based on large estates and the free enterprise of the investors. The settlement at St. Mary's was made in the spring of 1634 by a company of gentlemen and servants. The settlers grounded their agriculture upon maize, whose advantages as a feed grain they recognized, and adopted tobacco as their money crop. Here, as in Virginia, the marks of disease appeared early. Mortality rates were high, for an adult of 20 could expect to live only to 44.5 years; laborers were overwhelmingly male. High deathrates and sexual imbalance conspired to prevent the colony from maintaining itself by natural increase until the 1690s. Again a marked contrast to New England, where men born in the colony lived 10–20 years longer than men in the Chesapeake, and where, after a harsh beginning, a colonial society of family groups in a cold climate proved almost explosively fertile. Success came to Maryland late and at a high cost; yet by the next century the success was real. Like Virginia, Maryland began to grow by natural increase.

Throughout the Chesapeake, rapid population growth became the rule in the eighteenth century. The widespread use of Peruvian bark made malaria more endurable. During the hard decades the mean age of marriage had fallen; it remained low long after lifespans had begun to increase. "They marry very young," noted John Lawson in about 1708 of the North Carolinians, many of whom were transplanted Virginians, "some at Thirteen or Fourteen; and She that stays till Twenty is reckoned a stale Maid, which is a very indifferent Character in that warm Country. The Women are very fruitful, most Houses being full of Little Ones." In asserting that "those who are born here, and in other Colonies, live to as great Ages as any of the Europeans" he was by then almost right.[16]

The accommodation of the population to disease worked from many directions. The resistant survived and multiplied; at the same time, the most virulent strains of microbes were destroying themselves along with their victims. The turmoil of settlement shifted from the Tidewater countryside to the interior, where land was still to be had. Stable communities offered less opportunity for vectors to breed; and any planters who penned their cattle in defiance of custom did themselves an unconscious favor, for *Anopheles* prefers animal blood to human and will feed on the latter if available. Epidemicity passed into endemicity save for diseases (smallpox, yellow fever) that could not maintain themselves, the latter for reasons of climate, the former because of the scattered population. The durable remnant of the population, deprived of the competition of their fellows, engrossed and began to deplete the land, and bred rapidly until rising numbers introduced more sedate norms. People lived longer, the middle-aged held tenaciously to property, the young married later, and the landless, finding little opportunity, betook themselves westward. A Darwinian irruption, when differential mortality acquired extraordinary influence, gave way to the normal Lamarckian course of civilization.

The two Carolinas received a similarly heavy impress from their strikingly different environments. South from Cape Henry in Virginia to Cape Fear in North Carolina, the coast is fronted by barrier islands, for the region lies in the sandy deposits of the continental shelf. North of Hatteras the inlets are so shallow that, even by the standards of the seventeenth century, only the smallest vessels could enter. South of Hatteras several admitted "Ships of Burden, such as Ocacock, Topsail-Inlet, and Cape Fair [Fear]." The Cape Fear River penetrated the interior but was obstructed by "arduous shoals and sand bars." Depending heavily on its neighbors to ship its products, and discriminated against by them, North Carolina found its routes to Europe costly and indirect.

The soil of the coastal region reflects its origin in the sea. In 1765 a visitor termed the land "very sandy and indifferent . . . sea shells [are] in plenty, which would seem to intimate that [a] great part of Carolina was risen from the sands, thrown up by the Sea to a Certain height and then obliged itself to retire."[17] Not until the 1690s was this region, where Raleigh had planted his first colonies more than a century before, permanently settled by Europeans.

The society that grew up behind the coastal barrens was marked by subsistence farming on small units, by cattle raising, and by the naval-stores industry that gave the people of the region their sobriquet of "Tarheels." American trees had long been marked as a source of pitch and timber by the homeland, with its wooden fleet and its rapidly disappearing forests. The spread of slavery in the colony (in 1732 about one-sixth of the population was black) reflected the laborious nature of such production, as well as that of tobacco, which was grown largely by yeoman farmers, each possessing a few slaves. Especially in the lower Cape Fear country, settlers of considerable wealth practiced large-scale, systematic exploitation of forests and soils, extracting naval and wood products in the winter when hands might otherwise be idle. Cattle were important, especially in the east, and settlers continued to make use of fire to keep open the woods and enlarge the meadows for grazing. Treatment of land was as casual and negligent as in early Virginia. Yet for all the revolutionary changes that attended the coming of the whites, the introduction of wheat, and the sale of tobacco, much remained that was similar to Indian times. Small farmers grew mixed crops for subsistence, half-feral cattle roamed the woods with the deer, and almost all men hunted.[18]

In cool defiance of the thesis that frontiersmen were necessarily destructive, Moravian settlers in the eighteenth century created a society around Salem whose semiurban form, peasant practices, and religious dynamic were utterly different from those of their neighbors, and broadly conservationist in their treatment of resources. Intensive farming and care of the soil brought surprising results which most travelers noted: "The moment I touched the boundary of the Moravians." wrote one, "I noticed a marked and most favorable change in the appearance of buildings and farms, and even the cattle seemed larger, and in better condition."[19]

With its complex patterns of settlement, its varied lifestyles, and its lack of an urban focus, North Carolina differed drastically from its neighbor to the south. From Cape Hatteras to Florida the future Rice Coast stretched low, marshy, and mostly infertile. Creeks, inlets, and bays formed the Sea Islands. Mineral wealth was largely absent, the rivers short and obstructed by bars, and the pine barrens were good sources of naval stores but of small use for farming. In its early decades South Carolina bore little resemblance to the classic staple-producing colony of later times. Cattle and pigs, turned loose in the woods, throve despite predation by wild animals. The colony exported great quantities of salt meat. Forest products—pitch, tar, turpentine, staves, clapboards—were also important, as from the beginning were deerskins. The colony's evolution was swift, however; slavery was brought by the Barbadian colonists, headright was introduced, and the success of rice soon gave a staple crop.

Relatively high, fertile areas of "good oak land" or "hickory land" attracted agriculture. Rice land, by contrast, came to mean cleared swamp.[20] Besides the landform and the staple, a prime difference between South Carolina and the colonies of the Chesapeake was its semiurban character. Established on Ashley River, Charles Town grew into one of the four largest urban centers in the colonies and became a political, economic, and cultural power dominating not only the Rice Coast but a vast hinterland, as well. Through it funneled crops and an array of wild products. The usual early experiments with tropical crops brought the usual failures. But on the uplands of the coastal plain, indigo was grown after 1751 under Parliamentary encouragement as a plantation crop. In 1756 some 200,000 pounds of the dye were shipped from Charles Town, though this was the highpoint in a mushroom industry that declined swiftly thereafter in the face of Asian competition. Besides rice and indigo, the city was the center of the trade in pelts, rather like Albany in the North, and through this trade brought major environmental changes to the inland as well as the coastal south.[21]

The colony of Georgia, first established as a buffer to protect South Carolina from the continuing danger posed by Spanish Florida, evolved from a utopian project to a conventional southern reality. Again in the eighteenth century, James Edward Oglethorpe repeated the climatic (or rather latitudinal) arguments that Raleigh and Hakluyt had advanced in the sixteenth. By a settlement on the Savannah River, England could be supplied with "Silk, Wine, Oil, Dyes, Drugs, and many other materials for manufacture, which she is obliged to purchase from Southern countries." Settlement in 1732 followed a company plan approved by the Crown, not so fanciful as earlier suggestions of a Margravate of Azilia, but with considerable idealism intact. The colony was intended, besides its military function, to "provide for poor People incapable of subsisting themselves at Home," and was to be free of slavery and spirituous liquors. The usual effort to establish the silk industry was made, not without a modest degree of success, though never enough to make it practical. But the projects of the founders did not long resist the attractions of black slavery, and of white freedom from restrictions on the ownership of large estates. The right to hold land in fee simple, won in 1748, was a great impetus to slaveholding and the growth of large plantations, yet it appeared wonderful even to those who opposed slavery. "The most lowly peasant," wrote a German settler three years later, "is an absolute freeholder in his house and on his land, and cannot complain in the least about any difficulty, oppression and violence, which is exactly the famous English liberty."[22]

By the latter part of the seventeenth century, the spread of black slavery, not only in Virginia but in the newer colonies to the south, had introduced a new human factor of incalculable importance into the southern environment. With the growth of slavery a new part of the human family entered the South, transforming the land by its labor, society by its lifelong bondage, and disease patterns by introducing the ills of Africa.

Few subjects are shrouded in deeper uncertainty than the changeover from white servitude to black slavery in the Chesapeake colonies. In the first half of the century some blacks were held already in a bondage indistinguishable from slavery, others in a condition like that of white bondsmen, treated with about equal brutality, and at the end of a period of indenture set free to join the quest for land, servants, and status. Yet during the 1660s the advantages of slavery gained the upper hand. Prices fell as the British Empire entered the trade extensively. As chattel slavery emerged and was defined in law, the slave added to his master's capital as well as his workforce. Stability in that force was an important advantage in a land where bondsmen, both by the terms of their indentures and by reason of their high mortality, came and went in undependable fashion. [23]

There were also solid physical advantages in black slavery. However malaria was first introduced into the Chesapeake region, the establishment of the disease placed a premium on workers who were resistant to it. Blacks possessed substantial inherited immunity to one of the benign forms and, through the sickle-cell trait, resistance to the malignant form, as well. But their resistance was strong to many diseases. Some planters believed that blacks were not subject to the period of sickness in newcomers that was called seasoning. Newly-arrived blacks may in fact have experienced a lower mortality than white newcomers in the early decades of the seventeenth century. By and large these blacks came from the Indies, were acculturated to English or Spanish norms, spoke western languages, and were sometimes baptized. They must also have been people of singular immunological sophistication by the time they reached North American shores.[24]

Sub-Saharan Africa was well supplied with diseases, and wherever malaria may have originated, it had apparently been longest entrenched there. The equatorial lowlands, source of the first black slaves taken by western nations, were severely afflicted also by yellow fever, filariasis, yaws, sleeping sickness, and other ills. Except for sleeping sickness, which could not spread for lack of the tsetse fly, all of these took root in the West Indies. On the slave ships and again on the West Indian plantations, blacks encountered, and some survived, contact with the white man's diseases.[25]

African ills spread rapidly in the South's friendly climate. The malignant form of malaria was established. Yellow fever entered the West Indies from Africa in the 1640s, reaching Charles Town in epidemic form at the end of the century. It reappeared erratically thereafter, reintroduced by the slavers and the sugar trade which may first have given it currency. Yaws showed up among slaves in Jamaica and spread widely in the South among whites, blacks, and Indians. In one region of North Carolina near the Dismal Swamp, its lesions were so common in the early eighteenth century that William Byrd II jocularly reported a supposed motion in the colonial assembly to deny preferment to any man who had a nose. The filarial roundworm, cause of elephantiasis, became endemic in the Charles Town area and remained so until the twentieth century. Hookworm disease, caused by a parasitic nematode, and dengue (breakbone fever) were African imports that remained limited to the South.[26] Blacks, like whites, were reservoirs of infection, maintaining and spreading diseases to which they themselves were resistant but by no means always immune.

One result was to provide a durable justification for slavery. As African diseases took hold, and especially as malaria became endemic and yellow fever a frequent visitor, the notion spread in the South that hard work in sickly areas such as rice growing could be performed by blacks alone. In 1739 a group of Georgians, pleading that slavery be introduced into their colony, launched an armory of arguments drawn from the supposed physical superiority of blacks. The heat of the Georgia sun was "insupportable" to white servants, causing "inflammatory fevers of various kinds both continued and intermittent," as well as fluxes, colics, "dry belly aches; tremors, vertigoes, palsies, and a long train of painful and lingering nervous distempers, which brought on to many cessation both from work and life." The ills mentioned suggest malaria, some form of dysentery, and lead poisoning, the latter possibly caused by lead-glazed utensils in which acid drinks such as lemonade or rum punch caused toxic salts to form. Blacks have indeed less resistance to cold than whites, and greater natural resistance to humid heat; but with adequate salt intake whites can labor in the tropics quite as well. The petition assumed the connection between climate and disease, assumed the greater resistance of blacks, randomly extended black resistance to malaria to other ills that are entirely or largely color-blind, and used the whole melange of observation, assumption, and error to prove the need of Georgia for black slaves.

"Rice can only be cultivated by Negroes," remarked a French visitor in 1799, as a matter of common knowledge. "Slavery . . . confirms the planter in his prejudice for rice, and the cultivation of rice, on the other hand, attaches him to slavery." The growth of climatic theories of disease could only help to confirm what self-interest already declared: that some of the most profitable forms of southern agriculture were of necessity, and for all time, Negroes' work.[27]

The spread of slavery caused steadily increasing imports of African as well as West Indian blacks, however. Africans were neither so acculturated to European ways nor so likely to be experienced in surviving European diseases. The long voyage and the brutal conditions on shipboard weakened many, not only by physical attrition, but by inducing despair. The result was a new crisis of transplantation not unlike that experienced by the first white colonists. Black death rates surged, a fact which the dispatch of newcomers to the newest and rawest frontier plantations did nothing to alleviate. Few women were imported, making replacement of the dead by natural increase impossible. Pneumonia, influenza, and a rapid, deadly form of turberculosis struck the new arrivals cruelly. The cold regions of North America may have been most dangerous to blacks, but even slaves in the warmest were not exempt, in part at least because the sickle-cell anemias predispose their victims to pneumococcal infections. In 1728 South Carolina was visited with "a sort of pestilential pleuritick feaver" that hit the blacks hardest, and in 1755 Alexander Garden, a Charles Town physician, noted that after heavy labor slaves were "Seized with dangerous Pleurisies and peripneumonies which soon rid them of Cruel Masters, of more Cruel Overseers, and End their Wretched Being here."

In the importation of African as of European disease, the experience of transplantation and conditions on shipboard both played crucial roles. Captives from diverse villages were gathered in the barracoons of the African coast, then packed tightly aboard the slavers. From the arrival of the coffles at the slave pens to the end of the "seasoning" period in America, about 90 percent died from all causes. At the American end of the Middle Passage the conditions of the slave ships were chiefly visible. As port physician of Charles Town, Garden inspected many slavers. "There are few ships that come here from Africa," he reported, "but have many of their Cargoes thrown overboard; some one-fourth, some one-third, some lost half; and I have seen some that have lost two-thirds of their slaves . . . I have never yet been on board one, that did not smell most offensive and noisome, what for Filth, putrid Air, putrid Dysenteries (which is their common Disorder) it is a wonder any escape with Life." This was before the "seasoning" began; whether deaths during that process resulted more often from the effects of life on the slavers or from diseases contracted in the New World remains unclear.[28]

Under these stresses, more blacks died in the American colonies than were born. It was after 1720 before black life stabilized enough for the population to resume its interrupted growth. The Creole children of African slaves adapted as the whites had and outlived their parents. The disease environment of the South, while kinder to tropical ills than the North, was far less so than equatorial Africa. Sleeping sickness did not take hold, endemic filariasis was confined to one small area, and yaws later spontaneously disappeared with the slave trade that sustained it. By mid-century the trade had also begun to decline, and by the 1770s importation of Africans into the Chesapeake region had dropped to insignificance. (Heavy imports into South Carolina, however, continued, save for a few years after the insurrectionary alarms of 1765, until the Revolution.) By the end of the century substantial black and white populations in the South had adapted to the requirements of American life and to each other. Their survival, at any rate, was not in doubt.

The environment the blacks and whites were engaged in making in turn fashioned them. Their own domesticates and parasites created a world and they must live with it. Their agriculture and their diseases interacted in a synergy that transformed society.

"Rice," says a recent historian of slavery, "was a hard taskmaster." Introduced deliberately into South Carolina before 1690 as a plantation crop, the plant, because it required flooding during part of its development, drew agriculture and human habitation alike toward the swamps. The settlers, many from the West Indies, had introduced black slavery for the naval-stores industry and readily adapted the institution to the growing of rice. Making the new crop demanded hard, disciplined labor in knee-deep water under a semitropical sun. The responsibility of the slave trade for the deadlier forms of malaria is indicated by the fact that until 1678–1679 the settlements on the Ashley River were quite healthy. By the 1680s, however, a mild form of malaria had become established, probably brought from England, for Barbados was free of the disease. Polluted wells spread typhoid. Then, needing abundant labor as they shifted from early diversified crops to rice, the colonists brought in slaves, first from the West Indies and from buccaneers, later directly from Africa. In 1684 the proprietors wrote that "We are by all people informed yt Charles towne is no healthy scituation and it hath no good Water in it and all people that come to the province and Landing there & the most falling sick it brings a Disreputation upon the whole Country."[29]

Despite the usual difficulty in distinguishing between malaria and other fevers on the basis of contemporary reports, it appears that both the malignant and the benign form of the disease were well established by the mid-eighteenth century. By that time even the Georgian hinterland was the residence of *"febres intermittentes,* particularly daily and three-day ones." But the rice paddies that stretched along the tidal streams from Cape Fear to the Altamaha River probably formed the most endemic area, where a night in the open during the June-to-October sickly season was held to be fatal to the unacclimated.[30]

Indeed, few situations could be imagined which would have been better for the disease and its local vector. In shallow, still water, pierced by the stalks of rice to which the larvae could attach themselves for breathing and feeding, *Anopheles quadrimaculatus* found a perfect nursery. At home too was the plasmodium, the parasitic protozoan of malaria which attacks the liver and red blood cells. Close at hand was an abundant supply of warm-blooded hosts. Conditions of heat and humidity favored multiplication in its second host, the mosquito. Malignant (falciparum) malaria could kill; the so-called benign forms condemned the victim to paroxysms of chills and fever, anemia, general debility, and frequent relapses.[31]

Malaria affected the course of slavery, its tone, and the differences observed between different regions. Students of American slavery have remarked on the importance of the resident versus the nonresident planter. The nonresident handed over land and slaves to an overseer, probably with effects adverse to both. But the presence or absence of the master affected the character of slave life in subtler ways, as well. With the master at a distance, blacks tended to develop a distinctive society, while his presence favored the growth of an Afro-American lifestyle based on assimilation. Chesapeake planters were usually residents on their plantations, while most of those on the Rice Coast fled to the towns or the pinelands during the malaria season. These patterns clearly had much to do with the nature of tobacco and rice culture, in particular with the dispersed smaller holdings of the former and the concentrated large holdings of the latter, compelled by the necessity to flood the plant. Additionally, in sheer numbers the black presence was massively greater in South Carolina—about half the population in 1710. But the black culture of the region also had much to do with the association between rice culture and malaria. Variations in the disease environment helped to shape complex social identities as well as more elemental facts of comfort or misery, life or death.

As for the Indian, the colonization of so many deadly ailments continued to wreak its customary havoc. Smallpox was the chief killer. Historian John Duffy has described in some detail its ravages in colonial times. So common in England during the eighteenth century that it became a childhood disease like measles, though with a substantial continuing mortality, smallpox remained an imported and hence an epidemic scourge in the New World. Here the scattered American population prevented the disease from establishing endemicity until the wars of mid-century. Most of the South was comparatively little affected because settlements were so few, though a major epidemic did occur in 1679–1680.

Charles Town, a port with a vast Indian trade, was, however, a focus of infection. In 1738 an epidemic there was transmitted to the Cherokees, reportedly causing the death of half the tribe. Overall the pattern of infection continued to feature the usual brushfire flaring and fading, as smallpox was repeatedly reintroduced from abroad and carried by trade or war to the interior. White, red, and black swapped the disease about, the movements of population through the backwoods helped to carry it from one colony to another. In 1700 Lawson found the Congeree Indians of South Carolina a "small people, having lost much of their former Numbers, by intestine Broils, but most by the Small-pox, which hath often visited them, sweeping away whole Towns. . . . Neither," he added, "do I know any Savages that have traded with the English but what have been great Losers by this Distemper." The Pemlicoes, nearly wiped out in

1696, were in 1710 confined to a single village; the Sewee Indians perished by smallpox and rum and, in a bizarre episode, by attempting to sail a fleet of canoes directly to England to trade. The Catawbas, once the largest of the eastern Siouxan tribes, declined rapidly after their first contact with whites; they suffered terribly from the epidemics of 1738 and 1759, losing about half their number in the latter year. After that they "ceased to play a prominent part in history."[32]

The whites did not escape. Soldiers accompanying Governor Henry Little of South Carolina to a treaty signing caught the disease from the Cherokees in 1760, and brought it back to Charles Town, where many were susceptible because no outbreak had occurred since 1738. The disease hit uncommonly hard; three-quarters of the town became ill, and about 9 percent died. Indians continued to be the heaviest sufferers, however. Processes of cultural decay, abetted by the influx of European trade goods, slowly brought the native peoples closer to the settler's image of savages. Indian towns shrank or were abandoned; ceramics became debased as craftsmen died. Smaller communities forsook complex pursuits for the necessities of subsistence. But it was the physical disappearance of the Indians, the emptying of glade and woodland that had once been full, that struck some colonists as a mystery explicable only in terms of the supernatural.

Men of scientific bent confined themselves to natural explanations. The naturalist Mark Catesby noted that the Indians by and large had "healthful constitutions . . . little acquainted with those diseases which are incident to Europeans," though he found them short-lived by reason of the hardships of their lifestyle. He believed that they had helped to cause their own demise by incessant wars, but he also believed that the Europeans, "by introducing the vices and distempers of the old world, have greatly contributed even to extinguish the race of these savages, who it is generally believed were at first four, if not six times as numerous as they now are." More mystical was the view of John Archdale, former governor of South Carolina. Grateful for the "thining" of the Indians, he felt that the direct interposition of God in the process had saved the English from incurring much guilt of Indian blood: "It pleases God to send, as I may say, an *Assyrian* angel to do it himself." Contradictorily he praised God's handiwork and lamented its consequences, for he found the Indians "generally very streight Bodied, and Comely in person, and quick of Apprehension." He called for their conversion and regretted the fact that more talk than effort had gone into such work to date. Nevertheless, he had no doubt of the divine intent; recounting "the great Mortality that fell upon the *Pemlicoe Indians,*" in 1696 by "a Consumption," and the destruction of another tribe by internecine war, he wrote "that it seemed to me as if God had an Intention speedily to plant an English Settlement thereabouts; which accordingly fell out in two or three Years."[33]

The Problem of Survival   43

Clearly, the decline of the Indians was paced by many factors, including intertribal war and emigration. Moreover, there appears to have been another kind of synergism at work in the process of warfare, with European weapons and practices, the lure of European goods and the demands of alliance with the whites steadily raising the level of destructiveness of native war. There were many senses in which Charles Darwin's observation "Wherever the European has trod, death seems to pursue the aboriginal," could be interpreted. It is unlikely that the proportional contributions of the many destructive forces will ever be precisely worked out, but it is possible to say that the transformation of the disease environment by white and black ills was a fundamental force.

The balances of white, red, and black shifted within the South, in precisely the same way in no two places. Strong interior tribes of Indians survived, still to play major roles in southern—and national and international—history. But in the coastal regions the natives of the land were of small account by the end of the colonial period. European and African exotics, the plantation system which bound them, and the Atlantic world they served in such different roles, ruled the seaboard of the South.[34]

3

The Uses of the Wild

ONE OF THE BIBLICAL METAPHORS IN WHICH OUR ANCESTORS DELIGHTED WAS especially apt: a new earth had been given into their hands, with its creatures, trees, and waters. They were not all-powerful over it, and they were, as time went on, increasingly restrained by revived notions of stewardship. Nevertheless, they could not have failed to be struck by the upending of values which resulted from the crossing of the sea. In England labor had been abundant and land dear, men common and forests dwindling, privilege strong and animals of the chase protected. Englishmen ventured into their New World with all the wrong assumptions—or else with ideas and values that were unconsciously prophetic of an overburdened American earth that was yet to come.

Late in the seventeenth century new fashions, for science and nature, for simplicity and knowledge, rolled out as a wave from the most artificial centers of Europe and washed gently up the Atlantic beaches of America. Reconciling the English heritage and taste with western reality was a complicated business, but southern settlers set about it with a will, pursuing survival, then profit, then fashion, according to their lights as Englishmen.

Even taking with the customary grain of salt the rhapsodies of colonial publicists, the abundance of wildlife in the Indians' South must have been remarkable. Considering only the coastal areas first settled, a wealth of species existed, some of which can now be found only farther westward, while some have been altogether destroyed. Migratory fowl ranged along the Atlantic flyway; anadromous and freshwater fish were abundant. Settlers ate sturgeon big as cod. Flights of parakeets were reported as far north as southern Virginia. Passenger pigeons were stupefyingly abundant, perhaps the most abundant species of bird in history; one scholar has estimated the total pre-Columbian numbers of all the great flocks at three to five billion. In the region of the southern Atlantic coastal states there were, at a conservative estimate, about 1.8 million turkeys. The catalog of major animals included deer, bear, bison, cougars, lynxes, bobcats, and beaver.[1]

The colonists arrived upon this scene, not as calculating destroyers, but as confused and rather ineffectual intruders upon a world that differed from any they had known. Sheer increase in human population was not, until the eighteenth century, a significant factor in the decline of wildlife. Indeed, it is by no means clear when the total post-Jamestown population equaled that of the Indians alone in presettlement days, in view of the losses to disease suffered by both red and white. The last quarter of the seventeenth century might be a reasonable suggestion. Heavy white settlement in a few areas (the Virginia peninsula and the Charles Town neighborhood, for example) no doubt resulted in locally severe destruction of habitat. But in the peninsula, Indian overhunting had already caused considerable wildlife losses before the whites arrived. In any case, settlements in the Chesapeake soon became scattered, and the scrub country created by tobacco culture provided cover for deer, many smaller creatures, and birds.

Colonial hunting, like farming, was marked by regression from English norms. There the state sought to preserve both habitat and favored species, to restrict hunting by a variety of regulations, and to confine its pleasures to privileged persons. "[F]ifty times the property [is required]," wrote the jurist William Blackstone, "to enable a man to kill a partridge, as to vote for a knight of the shire." In the eighteenth century, English law still imposed heavy fines, imprisonment, or transportation for poaching.[2] Thus conservation was linked to injustice, and Blackstone wrote into the *Commentaries* his own contempt for the "little Nimrod[s]" of England who denied the poor an important source of food in order to insure sport for themselves. Americans were happy to agree with their great teacher in matters of law. With exceptions—notably in the South, where men of property sometimes claimed special hunting rights—New World practice tended toward a doctrine of free taking. This was democracy at work, reshaping the law to meet the new conditions of abundant supply and an immature class structure that characterized the colonies.[3] Yet Blackstone himself, in opposing private privilege, held that government could set the rules for the taking of game. This too persisted and passed into American law.

Colonists of the first generation retained some basic idea of the necessity for and practice of conservation. John Smith expressed surprise at the Indian habit of indiscriminate killing. After the 1622 massacre he argued that the destruction of the Indians would be the saving of the beasts: "Besides, the Deere, Turkies, and other Beasts and Fowles will exceedingly increase if we beat the Savages out of the Countrey, for at all times of the yeere they never spare Male nor Female, old nor young, egges nor birds, fat nor leane, in season or out of season with them, all is one."[4] Indeed, remains in Indian middens demonstrate that they ate both sexes and all ages of deer. But the settlers' custom of limited hunting soon decayed in the face of their own needs and the American abundance. They hunted to supplement their tables, for sport, and for the market created by the trade in skins; they drew the Indians into the trade; and they combined (as did the Indians) the methods of the red man, especially fire hunting, with the weapons and purposes of the white to form a most destructive combination.

Among the animals most hunted was the turkey, which the colonists knew in two forms, domestic and wild. The domestics had been found in Indian villages by the Spanish and transported to Europe, where as an exotic they acquired in England one of the contemporary names for anything foreign ("guinea" and "muscovy" were two others). Settlers found the wild birds larger and by common agreement more beautiful; hybrids between the wild and tame were larger and hardier than their domestic forebears. Hunters learned that the wild turkey was wary but foolish, quick but easily confused, and much given to venery. It survived four-footed predators with success, which did not prevent blame from falling upon wolves or foxes when the numbers of the turkey were seen to decline. The bird was exceedingly abundant in early colonial times. In 1612 in Virginia William Strachey saw "a great store" of turkeys, forty in a flock; in 1700 Lawson and his companions in Carolina saw flocks "containing several hundred in a Gang," and ate so much of the bird that they became "Cloy'd with Turkey" and gratefully ate "a Possum" instead.[5]

A source of durable wonder was the passenger pigeon, then a staple of dinner table and campfire. The pigeon was an elegant bird, about sixteen inches long, extremely rapid in flight, and possessed of a "red and sparkling eye." William Strachey was not the last to feel that he might lose "the credit of my relation . . . yf I should express what extended flocks, and how manie thousands in one flock, I have seen in one daie." In the next century the settlers of Louisiana had similar feelings. Unconsciously echoing Strachey's image of the flocks as "like so many thickned Clowdes," Antoine Simon Le Page du Pratz in 1758 declared, "I do not fear exaggerating when I assert that they sometimes darken the sun." John Lawson found small Indian villages that had on hand "more than one hundred gallons of Pigeon's Oil or Fat; they using it with Pulse or Bread, as we do Butter." And—again the same image, this time about 1700 in Carolina—"the Flocks, as they pass by, in great measure, obstruct the Light of the day."[6]

A special and curious case was the beaver, about which legends gathered almost as thick as about snakes, raccoons, and opossums. The human-like activity of building dams brought an urge to see more human ways in the animal, and beaver-captains were reported ordering about their construction crews and, of course, performing masonry work with their tails. The real impact of the beaver on its surroundings forms an interesting subject for speculation. The animal was widespread in the colonial South, despite the trapping that went on for the beaver hat trade. Lawson wrote that "Bevers are very numerous in Carolina, there being abundance of their Dams in all Parts of the Country where I have traveled." A few years later William Byrd II found that beaver dams had flooded extensive areas along the North Carolina-Virginia border, a region which he also found marshy, troubled by swarming mosquitoes, and beset by malaria. The transformation of flowing streams into still pools and the spreading of water over low-lying land which resulted could only have been welcomed by mosquitoes, including the anophelines. Again, flooding must have impeded the spread of fire, moderated streamflow, and retarded erosion. But swamps existed for many reasons besides beaver dams (Byrd was soon bogged in the fringes of the Great Dismal), and the impact of this interesting creature on the ecology of the region remains problematical.[7]

The prime game animal of the South, the foundation of the peltries, and the chief prey of the Indians was the deer. In pre-Columbian times all species of deer may have numbered some fifty million north of Mexico. If the present is a reliable guide to the past, at least a dozen subspecies of the white-tailed deer inhabited the South, the most important east of the Mississippi being the Virginia deer. They were long-established natives: in many details of their anatomy the white-tails are "so different from any of the Old World genera as to imply descent from a long line of American ancestry." Though graceful, elegant, and usually timid, the deer had proved to be able survivors. As a result of selective pressures imposed by an array of predators, they exhibited a well-developed instinct of fear, and even fawns lacking experience of dog or man used many evasive tactics when confronted by danger. The deer were prolific as well as elusive, increasing their numbers when given minimal protection during the breeding season.[8]

Because the fields and orchards were attractive to many species, killing by trap and gun, dogs and fire, was not all subsistence or sport. In the colonists' view much hunting was necessary self-protection. Bears and squirrels descended on the corn, as did the crows—"great Enemies to Corn-Fields [that] cry and build almost like Rooks." In North Carolina ripening apples were visited by "numerous flights of parakeets that bite all the fruit to pieces in a moment. . . . The havoc they make is sometimes so great that whole orchards are laid waste." The killing of predators may well have encouraged the plant- and nut-eating creatures, but their depredations were basically the American abundance working against the settlers instead of for them—so many wild bellies to be filled! The consequence was bounty hunting.[9] In 1700 South Carolina put a bounty on "small Black Birds and Rice Birds." In August 1734 Virginia required every tithable resident of the Northern Neck and Eastern Shore to turn in to a justice of the peace three crows' heads or squirrels' scalps. The reward was a tax credit of twenty-one pounds of tobacco for each head or scalp, and there was a penalty for failure to comply. In 1728 Maryland enacted similar laws, and again in 1758.[10]

Predators fell under a similar ban. Colonists recognized that such animals posed little danger for themselves; long experienced in man's ways and powers, even the largest creatures took to concealment or flight. But the importance of hogs and cattle, and the practice of turning them unprotected into the woods, gave a cast of legitimacy to the hatred of predators. In fact, the colonists themselves were spreading a feast before the carnivores, not only in the form of living cattle, but in the carcasses of hundreds of thousands of deer. The result was an explosion in the wolf population and a great likelihood that the creatures were living more by scavenging, which was easy, than by predation, which was not.

In these circumstances, the legislation against the wolf is curious and grim testimony to a persistent human blindness. Having set off the increase of wolves, the colonists spent a great deal of tobacco hunting them down. In 1652 Virginia levied a tax on horses to pay bounties on wolfheads, only to repeal it three years later. In 1669 a quota of heads was imposed on tributary Indians but repealed the next year as unsuccessful. In 1691 a bounty of 300 pounds of tobacco was offered for every wolf trapped, and 200 for every one shot; later 100 pounds was offered to Indians. The laws were reenacted in 1705. In 1720 the bounty was made a straight 200 pounds of tobacco for killing a wolf by any means, and in 1748 it was reduced further. Found too low to encourage hunting, it was raised again in 1765 and continued in 1766. Six years later at least a modicum of sense appeared in the legislation, as persons guilty of killing deer and leaving the carcasses were denied the bounty.[11]

Maryland enacted similar laws in 1728, 1758, 1788, 1790, 1792, 1797, and 1798. Since complaints were heard that residents of Virginia and Pennsylvania were bringing in wolves to collect the reward, justices of the peace cut out the tongues and cropped the ears of the wolfheads in order to prevent unscrupulous persons from selling them again. In 1695 South Carolina had attempted to use its Indians in the campaign. Affirming that the tribes west and south of Cape Fear had been given clothing and tools, as well as protection from their enemies, "in consideration whereof they have not hitherto been any ways useful or serviceable," the colony required every hunter to bring in annually one wolfskin, one tigerskin, one bearskin, or two catskins (probably lynx or bobcat), on pain of being publicly whipped. The frequent reenactments of such laws suggest that wolves maintained a noticeable presence, at least along the frontier, until the nineteenth century.[12]

The settlers did not, however, only stimulate and/or attack existing wild animals. They also contributed new ones, as some of their forest-ranging stock went wholly feral. Lawson noted in South Carolina that the "Cattel" were "very wild and the Hogs very lean"; by the same period (about 1700) many wild horses roamed the Virginia woods. These animals were small but tough and fleet of foot, and pursuing them had become a sport. Beverly's "hogs swarm like Vermine upon the Earth" suggests not only great numbers but a species gone wild. Such animals were a favorite target for hunters, and Virginia law required the taker to prove that his slaughtered porkers had been fair game.[13]

Clearly the colonists' effect on wildlife was complex. While it is reasonable to suppose that the total number of wild animals declined in settled regions, the population of certain species climbed as the colonists provided new sources of food or destroyed old enemies. The settlers hunted indiscriminately, set off population explosions by inadvertence, and tried to deal with the consequences by bounty hunting. They did not destroy species, but they did unbalance local ecologies in erratic fashion.

As to what most colonists thought of their hunting, little doubt exists. It was more than part of their subsistence or their lives. It was also part of their freedom, their distinctiveness, their dawning character as Americans. The wonderful freedom to hunt without any man's permission seemed to them not the least of their liberties:

Here Propriety hath a large Scope, there being no strict Laws to bind our Privileges. A Quest after Game being as freely and peremptorily enjoyed by the meanest Planter [settler] as he that is highest in Dignity, or wealthiest in the Province. Deer and other Game that are naturally wild, being not immured, or preserved within Boundaries, to satisfy the Appetite of the Rich alone. A poor Laborer that is Master of his Gun, &c., hath as good a Claim to have continued Coarses of Delicacies crouded upon his Table, as he that is Master of a great Purse.[14]

Already efforts to preserve game on large estates had begun to seem downright un-American. Some revolutionary constitutions would specify the liberty to fowl and hunt as among the rights of free-born Americans.[15]

Far more serious in its effects than subsistence, sport, or even bounty hunting—so serious as to alarm the colonists themselves—was the trade in skins. The seventeenth century used leather for a great many of the purposes for which later centuries used metal or synthetics, as well as for shoes and winter clothing. The beaver hat was ubiquitous. In the early days at Jamestown corn had been the Indian product most in demand among the settlers. As the danger of starvation lessened, the focus shifted to deerskins and fur. Here were other surplus commodities by which the tribes could get access, not merely to the trivia of beads and copper, but to a variety of artifacts from guns and knives to iron pots. In all-too familiar fashion such objects quickly changed from luxuries to necessities.

The sourthern fur trade grew despite the competitive advantage of the longer, thicker furs of animals in colder climates. From the beginning, also, the trade implied giving the Indians what they wanted in return (guns were a favored item) and as a result Virginia intermittently legislated against it, or at least tried to license and restrict it. Nevertheless, from an early date pelts became a currency between the two peoples, and as early as 1620 there were said to be some 100 fur traders in the Chesapeake area.

Yet for climatic reasons 80 percent of Britain's fur imports came from New York, New England, Canada, and Hudson's Bay. More important to the South than fur was the trade in leather. Falling tobacco prices in the late seventeenth and early eighteenth centuries made deerskins increasingly important; enthusiastically developed by Governor Sir William Berkeley, by 1674 the trade had become a lucrative supplement to tobacco as a colonial export. Decline in the local deer herds meanwhile sent traders west and south. Frontier posts, established to protect traders and settlers alike, became important to the westward expansion of the colony. Traders, however, spread their activities far beyond any possible limits of Virginia, penetrating the Gulf South and the Mississippi valley. The trade boomed, with the "country south of Virginia probably surpassing any other part of America."[16] In the next century it fell rapidly into the hands of Charles Town traders. South Carolina, with Georgia and West Florida, now received the lion's share of the trade. Between 1731 and 1765 Charles Town is said to have exported a minimum of 150,000 pounds of skins a year; from 1765 to 1773 Georgia exports were steady at 200,000 pounds annually.[17]

As historians have pointed out, this great slaughter deserves comparison in numbers (though not in its effect upon the species) with the later butchery of the buffalo on the great plains. It was made possible not only by arming the Indians with guns, but by fundamentally transforming their relationship with the deer from predator and prey to vendor and commodity. In 1728 Byrd remarked that the Nottoway Indians of Virginia, both for hunting and for war, "use nothing but firearms, which they purchase from the English for skins. Bows and arrows are grown into disuse, except only amongst their boys."[18] But though newly armed, the Indians' traditional practice of indiscriminate killing survived. As in days past, the woods were fired, usually when October had made them dry, and "the animals are drove by the raging fire and smoak, and being hemm'd in are destroyed in great numbers by [the Indians'] guns." In earlier days the meat had been as valuable as the pelt; hunger had regulated slaughter. Now only the pelt was valuable, needed to satisfy wants that grew as goods fed them, and that in consequence were essentially boundless.[19]

The tribes had been drawn into the Atlantic world with a vengeance. The transformation of war accompanied that of hunting, as the Indians, competing for the trade in skins, urged on by traders, and fighting beside white allies, learned more of the ways of European war. The cruel but ritualized conflicts of the past—fought for honor or vengeance, usually with small numbers of victims—tended to give way to struggles on the Old World model. More disastrous still was Indian success in achieving a lively trade. The influx of goods brought wealth, accompanied by dependence, and followed inevitably by a loss of self-determination and cultural distinctiveness. Native crafts decayed before European competition. Social ties based on the respect felt by the young for elders who were masters of traditional crafts grew weak. Regressive under the impact of disease, Indian culture suffered further erosion because its bearers went whoring, not after strange gods, but after strange goods.[20]

For a time the influx of weapons strengthened the great tribes, notably the Creeks and the Cherokees, who proved apt students of *realpolitik* as well as formidable warriors. But political decline followed cultural decay when the ouster of the French deprived the Indians of their ability to balance between the European powers. Weapons and rum then made them hopelessly dependent on the English alone.

The environmental impact of the skin and fur trades throughout the eastern part of the North American continent was formidable, leading to the near-extermination of favored species, especially in the North. Yet survival no less than slaughter keynoted the time. Compared to the great animals destroyed during the Pleistocene, the beasts of the time were more experienced in the ways of men, probably more wary, generally smaller, and perhaps better camouflaged. Though large and solitary animals tended to disappear from the settled areas, no species is known to have become extinct in the colonial South. Even the deer endured.[21]

Like the animals they sheltered, the southern woods underwent great changes. As they had with wildlife, English settlers came equipped with the assumptions and habits of scarcity and with legal traditions that emphasized sovereign rights over forests. Excluded from the Indians' clearings, the settlers confronted an overwhelming abundance which, however, was marked from the beginning by certain paradoxes. Trees obstructing land they wished to plant were mere nuisances, but some varieties, such as black walnut, early proved to have value as exports. The lack of roads meant that trees could not be moved except by water; hence, amid a seemingly endless forest, colonists faced the problem that trees which were both valuable and movable were comparatively few. Besides, trees served a variety of functions. They marked different varieties of land: obviously superior to that overgrown with conifers was the land where "big oaks, nut and other leaf trees grow, which is really the good fertile land which bears without fertilizer." Trees were the universal building material, the source of ships' timbers and of naval stores. Nut trees were an important source of food, the sugar maple was a treasure, the mulberry promised food for silkworms, and the wealth of flowering species rejoiced the eye. Indeed, from fuel to dyestuff and from soap to coffins, the array of forest products was all but endless.[22]

Though all the colonies exported fine wood, the naval-stores industry was especially associated with the Carolinas. Labor-intensive even by the standards of the times, the production of oakum, turpentine, tar, and pitch from resin helped to concentrate large numbers of slaves in the seaboard counties. The industry was early drawn into the imperial system; in 1705 naval stores were made an enumerated article, and premiums were paid on masts, hemp, bowsprits, spars, and resin products. North Carolina became a lumbering state, not only supplying ships but turning out shingles, staves, and sawn lumber. By the Revolutionary era, several dozen large sawmills were at work in the Cape Fear valley.[23]

Attempts to regulate cutting came from the Crown. The rapid growth of the British navy under the Protectorate and the Restoration coincided with a growing timber shortage in England. The difficulties of obtaining the large timbers needed for masts from the Baltic countries turned the government's eyes increasingly to the American colonies. A shipment of masts had gone to England from Virginia in 1609, but by the 1630s the trade centered in the northern colonies. White pine masts in particular preoccupied the Crown, and the concern was reflected in the Broad Arrow policy which extended the king's forest, marked by the inverted-V blazon, to North America. Originally limited to the Massachusetts Bay (1691), the policy of reserving suitable timber was extended in 1711 to all of New England, New York, and New Jersey. In 1729 the Broad Arrow came at last to the South, as to all of North America which was or might come under the control of England. In the South the most useful wood was live oak from the Carolinas, Georgia, and Florida, which provided sternposts and catheads to support anchors. The legislation, like so much mercantilist law, proved to be more annoying than effective. Colonists tended to ship their timber where they could find markets for it, and resented not only the reservation of the best trees, but the fact that violations of the Broad Arrow were tried in admiralty courts in which there were no juries. As in the case of wildlife, the freedom to exploit the forests at will was a practical liberty, and its denial a source of irritation between colonists and homeland.[24]

Forests were cut less for trade, however, than for domestic needs, and much was wasted. Most trees were destroyed merely to get them out of the way; much wood was burned, left to rot, or thrown into streams, whose obstruction provoked a small body of colonial legislation. As trees fell, fences rose. Much less work would have been required to pen cattle and hogs in than to try to fence them out, and in fact progressive farmers (of whom there were increasing numbers in the eighteenth century) did pen their stock. But if the animals were penned they must be fed, while if turned loose they fed themselves—after a fashion. Hence southern colonies (and states) required the enclosure of fields, not of stock.[25] The labor committed to fences, to say nothing of houses, barns, outbuildings, and so forth, and the danger to wandering livestock then caused colonial governments to legislate against the burning of woods. But fires were so easy a way to clear undergrowth, encourage grass, and destroy pests— and to aid hunting, though this practice was outlawed as well—that the woods continued to burn as before.

It is, by the way, notable as a regional trait that all legislation against burning was passed in the South only after the Revolution, though every colony except Maryland, Virginia, South Carolina, and Georgia had enacted such laws before that time. The South, with its extensive cattle industry, had long considered forests to be in some measure common property. The colonies south of the Potomac also failed to pass laws against trespass until long after the North and Maryland had done so.[26]

Despite the fact that large-scale deforestation was in its infancy, southern forests by the time of the Revolution had been modified by the rapid growth of clearings, by market production of timber, by selective cutting of commercially important varieties, and by the conversion of some coastal pine barrens into factories of naval stores. A careful student of a colony where much of this activity was centered has written: "By 1775 the forests of North Carolina were much altered and substantially diminished. . . . Succeeding generations were able to . . . continue the tradition of commercial forestry that the colonists had established. But the woodland they encountered was nowhere like that which the colonists had found, and in certain localities there was probably no resemblance at all."[27]

Yet even in the midst of revolution much abides. The forests had long been shaped by their human inhabitants, and the brief centuries of the colonial era could not transform them as widely as the millenia of Indian occupation had. Many of the processes at work in the colonial South were extensions of Indian practice. The new, emerging ecology of the rural South featured native and exotic species mingling amid farms and lingering woods. Near the coast the wild contracted, and beyond the settlements it resumed its sway. Between stretched a mixed region, part wild, part rural haven, not without profound echoes of the southern past. Johann Martin Bolzius, leader of the Salzburgers at Ebenezer in Georgia, described such a backwoods region in 1751 for his countrymen who might be contemplating emigration.

Two centuries after white penetration had begun, the animals which surrounded the German farmers were a mixture of natives and imported domesticates. "Neither in Carolina nor in Georgia," he wrote, "are there lions, panthers, and tigers. What are called tigers are only lynxes, which do harm to the cattle, pigs, and fowl. There are plenty of bears and wolves. But they are very timid, and flee when they see a man." Of alligators there were "a very large number," but they were not to be compared to Egyptian crocodiles. "They do harm to pigs, geese, ducks, and dogs that get too close to them," he acknowledged, but added: "There is, by the way, no reason to be afraid of them."

Agriculture showed clear resemblances to Indian times. Fire was still the main reliance for clearing. After the brush had been cut and the trees felled, "when one observes that all branches and bushes are quite dry, one puts fires to them and lets them burn up. Since the land is full of dry leaves, the fire spreads far and wide and burns grass, and everything it finds." To plant Indian corn, beans, and pumpkins, "one just makes holes in the earth . . . and after that prevents the grasses from growing. Potatoes or sweet roots one plants in round or elongated artificial mounds just like graves." The planters kept oxen and pigs, slaughtering them in autumn, and kept a long roster of other domesticates—"chickens, geese, ducks, Calcutta chickens, sheep, lambs, calves." People who lived near mills baked their own bread, while those who lived far away baked "certain cakes from flour of Indian corn or wheat on boards at the hearth." As in former times, however, "whoever has time and skill goes deer hunting, or shoots wild Indian or Calcutta chickens, wild ducks, etc."[28] This was the new South and it was pressing heavily against the old. Yet in much of its agriculture, its corn-based diet, its methods of clearing, its dependence on subsistence hunting, it was a world that the Indians of DeSoto's time would not have found absolutely strange.

The colonial period cannot be termed a classic era of conservation. Even when laws were enacted to protect trees or animals or fish, they were usually provoked by pain in the pocketbook and were of little effect. Yet such laws did keep alive the concept of the sovereign as disposer of resources, providing an alternative to the view that the natural world was a commodity and nothing more. A revival of interest in legal efforts to conserve reflected changes in the landscape. Both in the backwoods, ravaged by deer hunters, and in the Tidewater lands of the Chesapeake, cut over and misused by tobacco culture, the myth of abundance got some astringent dashes of scarcity in the eighteenth century.

Especially interesting was the effort of the southern colonies to legislate against the slaughter of the deer. In Virginia the shift in attitudes from the early days was clear-cut. In September 1632 an act forbidding the killing of wild hogs except on the hunter's own land went on to declare that anyone might kill deer or other wild animals, in order that the inhabitants of the colony might be trained in the use of guns, the Indians kept away, and wolves and "other vermine" destroyed.[29] By 1699 two generations of killing—and the growth of the skin trade into a major industry—had brought a different view. Now Virginia lawmakers found that "the Deer of his majestyes colony and dominion is very much destroyed by the unseasonable kiling them . . . and of Does big with young to the great detriment of the inhabitants of this his majestyes colony and dominion without bringing any considerable benefitt to those that kill them." A closed season was established from February 1 to July 31, when killing deer was forbidden under penalty of a fine of 500 pounds of tobacco, or 30 lashes for slaves.[30]

Thereafter the legislation was steadily broadened. In 1705 an act extended the closed season. In 1738 some disorderly persons were declared to be disobeying the law by killing deer for their skins and leaving the flesh to rot, attracting wolves. Some contributed to the destruction of the deer by letting their hounds run free in the woods, some by fire hunting, which was declared "destructive to the breed of deer" as well as to young timber and food for cattle. A closed season was again imposed, plus a longer one for does and fawns, under penalty of a twenty-shilling fine, or twenty lashes for slaves. Fines were imposed for letting beagles run loose and for fire hunting, and constables were given a limited right of search. In 1761 continued killing was noted, the fine increased, and whipping prescribed for any guilty party who could not or would not pay. Nine years later the lawmakers expressed fear that the whole breed of deer would soon be destroyed through the continued use of dogs for hunting; a fine of fifty shillings and cost was imposed for each offense. No deer at all were to be killed until 1776, and the grand jury was ordered to make inquiry, the better to discover offenses.[31] Clearly, the concepts of an endangered species and of extinction had entered both the minds of the Virginia lawmakers and the letter of the law.

Similarly in 1729 and again in 1730 Maryland enacted laws for the preservation of the breed of wild deer. An off-season was enacted and persons in possession of deermeat during the closed season were to be presumed guilty unless they could prove their innocence. Indians were exempted from the season but forbidden to kill deer for sale. Later acts to the same effect were passed in 1773, 1785, 1789, 1795, and 1800. The last law recognized that certain species of deer were likely to become extinct, and made killing them illegal for three years on pain of a £30 fine.[32] North Carolina established a closed season in 1738 and extended the law in 1745. In 1768 the colony, faced with continuing abuses caused by the skin trade, reverted to English practice by denying the right to hunt deer to anyone not having a 100-acre freehold or 10,000 cornhills, under penalty of a £10 fine and forfeiture of his gun. In 1774 night hunting was forbidden, and five years later the lawmakers declared that anyone called to give evidence who refused to testify might be sent to jail until he did. In 1769, meantime, South Carolina passed a similar law, citing the fact that too many persons were killing deer for their skins and leaving the flesh to rot, drawing beasts of prey, which threatened domestic animals. In 1785 the lawmakers cited the unnecessary destruction of deer by fire hunting and hunting at night; they imposed a fine of £20 sterling or three months' imprisonment, and for slaves whatever the magistrate might direct short of the loss of life or limb. Three years later a state at war declared that anyone convicted of fire hunting should be deemed a vagrant and enlisted in the militia—this as part of an "act for completing the quota of troops to be raised by this state for the Continental service."[33]

Somewhat similar in motive were the efforts made in the colonies and their successor states to regulate fishing. In 1680 Virginia outlawed gigs and "harping irons" during an off-season; in 1726 South Carolina forbade the poisoning of fish; Maryland, seeking "preservation of the breed of fish", outlawed fishing by weirs or dams in the Patuxent and Susquehanna rivers in 1766, and passed other regulatory laws in 1771, 1777, 1784, 1791, 1796, and 1798. Here were attempts to preserve creatures of great commercial importance which also supplied the tables and provided sport for colonial Izaak Waltons. As in the case of deer, the frequent reenactment of such laws suggests their ineffectiveness; they may indeed have become simply a rhythmic pietistic response to periodic complaints. But at any rate it is clear that the myth of abundance, at least in the over-hunted, over-fished seaboard regions, became badly tarnished during the eighteenth century. By 1773 every colony, even youthful Georgia, had enacted game legislation, most of it directed toward saving the deer.[34]

A few protests were heard against the misuse of domestic animals, as well. While running the boundary line between Virginia and its southern neighbor, William Byrd II professed himself astonished that the North Carolinians grew corn only for themselves and not for their animals. "Both cattle and hogs," he reported, "ramble in the neighboring marshes and swamps, where they maintain themselves the whole winter long and are not fetched home till the spring. Thus these indolent wretches [the North Carolinians], during one half of the year, lose the advantage of the milk of their cattle, as well as their dung, and many of the poor creatures perish in the mire, into the bargain, by this ill management." Such was the southern norm in new areas; but the Tidewater planters could no longer afford it.³⁵

Declining crop yields, rising population, and too many slaves created among the same people an interest in soil conservation. By all accounts the Tidewater planters rested much of their comfortable lifestyle upon their speculative investments in western lands. Yet pressure was considerable to diversify crops and improve the land on which they lived, partly because new land could only be had at great distances, partly in view of erratic tobacco prices and the presence of a large slave population that was becoming hard to employ profitably. Like George Washington, many came to realize that "much ground has been scratched over and none cultivated or improved as it ought to have been." The list of major agricultural reformers came to include many made famous by the Revolution, besides Washington—Thomas Jefferson, James Madison, and John Taylor were notable—but also lesser figures of Virginia and Maryland, such as Israel Janney, John Hinns, and John Singleton. Landon Carter, one of the wealthiest Virginia planters, recorded in his diary not only his private miseries, but his daily round as a careful farmer. In running his estates, Carter kept detailed records, read and wrote on agricultural topics, grew a variety of crops besides tobacco (wheat, maize, barley, rye, vegetables), penned his stock, used manure, studied the pests that attacked his crops, and experimented with various methods of control. He studied Le Page du Pratz's *History of Louisiana* with an agriculturist's eye and speculated on the nature of the wheat rust described there. Carter's scientific writings won him election to the American Philosophical Society in 1769, and a few years later to the Virginia Society for the Promotion of Useful Knowledge. In the effort to apply system and thought to farming, this troubled, difficult man stood with the best of his contemporaries.³⁶

Jefferson, during the domestic episodes of a life given to politics, worked with his land in Albemarle County, trying out crop rotations and deep and contour plowing.³⁷ Yet the work of the planter reformers enjoyed little lasting success, nor does it appear to have been widely imitated by their neighbors. The politicians among them were too much absorbed by other matters, farming mostly through overseers in the interludes of larger careers. Their experiments remained hit-or-miss affairs, for to design a truly satisfactory experiment is a technique, or rather an art, not easy to learn, much less to invent. Many discovered panaceas—marl, green manure, common salt, or whatever—which worked, or appeared to work, in one area, and then advertised it as a cure-all. Yet they had little precise knowledge of how particular fertilizers acted upon particular soils, and since the method that worked in one field might fail in another, even famous men found it hard to provoke imitation from others. Nor could soil conditions be separated from the other problems that might conspire to make a farmer succeed or fail—market conditions, credit, transport.[38]

Attempts to save the beasts and restore the land showed very limited results. Yet a consciousness of the need to save and restore existed, drawing its strength from the most pragmatic sources and enlisting able men—indeed some of the ablest the South has produced. The perception, the attempt, and the failure formed the practical side of the soil conservation story in the South of the eighteenth century.

The growth of knowledge was another matter, by no means uniform, but marked by real achievements that brought the southern wilderness into the consciousness as well as the marketplace of the Atlantic world. The movement toward greater knowledge of American nature was launched by clever visitors or immigrants who carried in their luggage the hobbyistic enthusiasms of seventeenth-century England's amateur botanists. Linked to American nationalism—what did so young a land have to glory in but its wilderness?—scientific and literary discoveries were forwarded by sophisticates: planters, townsmen, and Europeans. A leader in this as in other things, Jefferson mounted a hot defense of American animals (including its people) against the "imputation of impotence" by Buffon, a curious episode mingling the ardors of a scientist, a politician, and a patriot.[39]

It is true that the most needed form of knowledge was not attained. Eighteenth-century medicine remained generally deplorable, and public health measures primitive at best. Merely to read the endless dosing that goes on in Landon Carter's *Diary* is painful. The bleeding and purging by which Dr. Alexander Garden of Charles Town "prepared" his patients for inoculation against smallpox makes their survival seem more wonderful than many things reported of Lourdes. If this was amateur and professional medicine at the bedside level, public policy was no better. Governments were in general little concerned with health, whose protection was considered an individual or at most a local problem, though they might establish quarantine systems at ports, erect pest-houses, and try to exclude sick strangers during epidemics. When disease had taken hold in a community, attempts were made to purify the air; in Quaker Philadelphia and Spanish New Orleans, hospitals were built. Charles Town, an important focus of disease, showed considerable medical leadership in the eighteenth century, however. In the 1738 epidemic, inoculation was used to good effect, and statistics were gathered which provided solid evidence of its effectiveness.[40]

The natural sciences did better. Embodying many of the trends of the time was William Byrd II, no scientist, but a patron and friend of investigators and a man both fashionable and aboundingly curious. Born in Virginia, he was educated in London, returning in 1705 as heir of Westover. In 1728, as a commissioner on the survey to fix the boundary between Virginia and North Carolina, he had ample opportunity to see and record his views of American nature in his journal and later *History of the Dividing Line*. When the book was published, he added to it a number of passages extolling natural beauty and simplicity, presumably to make it more acceptable to English tastes. Privately he seems to have loved the rural rather than the wild, musing that a lovely valley "wanted nothing but Cattle grazing in the Meadow, and Sheep and Goats feeding on the Hill, to make it a Compleat Rural LANDSCAPE." Favoring the cultivated world in many senses, he patronized men who contributed to the science of the day, for science was a growing reality as well as a fad.[41]

Part of the interest in things natural sprang from the fashion of gardens among the wealthy in England, and the one-upmanship implied in "rare plants and extensive collections."[42] Before his death in 1692, John Banister, a young clergyman who was protégé of a botanizing bishop of London, sent home from an American journey specimens of plants and animals, and began a natural history of Virginia. Mark Catesby was an Englishman who acquired his first knowledge of America in 1712 while visiting his sister and brother-in-law at Williamsburg. With William Byrd II he spent seven indolent but not wasted years in Virginia. Returning from a trip home in 1722, now under the patronage of Francis Nicholson, first royal governor of South Carolina and a member of the Royal Society, he collected specimens through Carolina, the Georgian borderlands, Spanish Florida, and later the Bahamas. He made watercolor sketches as well, often working from life, and used native plants on occasion as settings for birds, animals, and fish.

Catesby's limitations were many. Tending to stylize his art, he was not always accurate; his white and brown curlews, for instance, were actually an adult and an immature white ibis. Yet he did unique work not surpassed for almost a century. One of his accomplishments (quite incidental to his main work) was to carry out with the aid of some slaves apparently the "earliest technical identification of the mammoth and even of a vertebrate fossil in America." He wrote that "At a place in *Carolina* called *Stono*, was dug out of the earth three or four teeth of a large animal, which by the concurring opinion of the *Negroes*, native *Africans*, that saw them, were the grinders of an Elephant. And in my opinion they could be no other; I having seen some of the like that are brought from *Africa*." In England, he prepared his great work, *The Natural History of Carolina, Florida, and the Bahama Islands*, which established him as the first American ornithologist. His work won fame and was widely purchased by the influential in England and America. His drawings, combining at their best much useful observation with the naive charm of primitive art, marked the beginning of a tradition in exploration, science, and painting that was to culminate in Audubon.[43]

No scientist, but an uncommon anecdotalist of the southern wild (among many other things) was John Lawson, a land speculator, explorer, and collector. His *Voyage to Carolina* (1709), a promotional work, spread much lively misinformation about southern animals along with its solid observations on Indian and colonial life. Here the curious could learn that the raccoon was a great drunkard, that the rattlesnake charmed its prey (a durable fancy), and that the male opossum had his "pizzle . . . placed retrograde and in time of coition turn[s] tail to tail." Lawson contributed to the development of the towns of Bath and New Bern, and introduced settlers from the Palatine into North Carolina. On an expedition to the backwoods he was taken prisoner by the Tuscarora Indians, who accused him of stealing Indian land, and by one account "stuck him full of fine small splinters of torch-wood like hog's bristles and so set them gradually afire."[44] He is believed to have died under torture.

The travels and work of John Bartram, the Quaker botanist of Philadelphia, also helped to fill in the portrait of the southern wild. In 1738 Bartram embarked on a 1,200-mile expedition through Virginia, assisted by the ever-curious Byrd. The *Description of East Florida with a Journal Kept by John Bartram of Philadelphia* recorded his expedition to the new possession of England in 1765. Throughout his works, his anecdotal style remained better disciplined by fact than those of his predecessors; though self-taught, he was a scientist of limited theoretical range but an excellent observer. Bartram's acquaintance, John Clayton of Gloucester County, in his *Flora Virginica*, produced the first Linnaean work on the plants of the New World. Clerk of Gloucester County, planter, an English-born provincial savant in the eighteenth-century mold, Clayton corresponded with Linnaeus, utilized his system of sexual classification, and adopted his binomial system of scientific names. Like Bartram, his connections to scientists and collectors throughout the colonies and abroad enabled him to participate in the intellectual universalism of the Atlantic world. Also like Bartram, he drew the impulse of his science from an abounding Christian piety that saw the hand of God in all things. As a practical matter, he derived the opportunity to pursue science from his status and prosperity as an owner of plantations and slaves.[45]

Dr. Alexander Garden also was a scientist of some attainments, though the gardenia, named for him by Linnaeus, was not his own discovery. Of Scotch birth, he studied medicine at Edinburgh under a botanizing pupil of the great Dutch physician, Hermann Boerhaave, himself a botanist and a patron of Linnaeus. Arriving at Charles Town in 1751, Garden found his professional opportunities rather too good in a land poor in doctors, rich in butcherly amateurs, and scourged by disease. In his own life he exhibited the situation of the physician inclined to science whose patients take time he could wish to use for other pursuits. "I am bound as if *manibus pedibusque* [hand and foot]," he wrote. "The most pitiable slave must be as regularly seen and attended to as the Governor.... From seven in the morning till nine at night, I cannot call half an hour my own." A good doctor of his time, he believed firmly in the physiological importance of climate and wrote feelingly of the Carolina weather to a British friend in 1759: "Were you to sweat out, for two or three summers, the finer parts of your good English blood and Animal spirits & have every Fibre & Nerve of your Body weakened, relaxed, enervated, & unbraced by a tedious Autumnal Intermittent, under a sultry, suffocating insufferable sun, you would then be made in some manner a judge of the reason of our want of taste or Fire. How different would your looks be? And how different would your Sentiments be? . . . instead of Fire & life of imagination, indifference and an gracefull despondency would overwhelm your mind."

Garden's own mind was not overwhelmed. He became a fixture and an ornament of the town. Like many American doctors he was friendlier to inoculation for smallpox than the English, and led the mass effort that took place during the epidemic of 1760. As a botanist he enriched Linnaeus's work, and his connections with Bartram, Benjamin Franklin, and others of similar interests was close. Intensely civic-minded, a man of the English-speaking urban South (when for practical purposes that meant Charles Town and little more), Garden achieved at some cost to his scientific attainments a rarely balanced life as both savant and practical man. To a great degree he answered his own complaint about South Carolina when he wondered "how there should be one place abounding with so many marks of divine wisdom and power, and not one rational eye to contemplate them."[46] Elected to the Royal Society in 1773, he became its vice-president in 1783 after he had returned to England.

During his trip to Florida, John Bartram left behind his son William, who wished to become a planter on the St. Johns River. Failing at that and at several subsequent jobs ("hard labour don't agree with him," his father complained to Garden, a family friend), William fled his debts to North Carolina. Later, however, he secured his father's help for a four-year journey through the Carolinas, Georgia, and the Floridas (1773–1777). The account he wrote in his *Travels* combined anecdote, scientific observation, and lush rhapsodizing upon nature. Sensibility was rampant. Noble redmen "dressed with simple elegance" paced his wilderness; its creatures were full of every sort of moral character, good and bad. Alligators were subtle and greedy and, for usually timid creatures, surprisingly determined to eat the naturalist. Yet Bartram could also witness in language both exact and poetic the takeoff of cranes: "They spread their light elastic sail; at first they move from the earth heavy and slow; they labor and beat the dense air; they form the line with wide extended wings, tip to tip; they all rise and fall together as one bird."[47] He could pause to note the curiosity of the coachwhip snake and the varying voices of frogs, and he seemed to feel an unfeigned fellowship with people, thunderstorms, birds, trees, and mountains which was (and is) very attractive.

His account, published in 1791, was treated with respect but little more in America. In England and France, however, the work caught the romantic imagination and ran through nine editions in a decade. Read by Coleridge and Wordsworth, Bartram's image of the wild passed directly into the enthusiastic nature-worship of those bookish men. In 1801 Coleridge wrote on the flyleaf of a copy of the *Travels,* "This is not a book of Travels, *properly speaking;* but a series of poems, chiefly descriptive, *occasioned* by the Objects, which the Traveller observed." Fair enough; and Coleridge proceeded to use Bartram as Bartram had used the southern woods. The image of waters upwelling in the wild, a real feature of Florida and a favorite subject of rhapsody by Bartram, found in a poet of the unconscious a powerful new significance. With verbal changes signalling the hand of a more powerful writer, Bartram's springs became Coleridge's fountains of Xanadu. It was a curious conjunction—the botanizing American ne'er-do-well bowing to immortality, and having immortality return the bow. Through Bartram the southern wilderness began to pass into literature, where it would take increasingly fervent root as the reality declined. A certain vision of that landscape, anyway: closely observed, yet with all its colors gorgeously inflated; the wild gone sublime and picturesque. Odder things have happened (but not many) than that the prickly Florida woods, transformed into a series of verbal aquatints, should become source material for *Kubla Khan.* Coleridge was not alone. Wordsworth used Bartram in some of his less happy poems. François-René de Chateaubriand, in his romance *Atala,* seized upon and developed his image of the southern wilderness and its noble savages. A connection was formed between ecstatic declamation and the southern woods and weather that only grew stronger with the passing years.[48]

For the future, the gifted migrants and native sons who colonized science and poetry in the region had done much. The southern land had entered the era of settlement as a realtor's Eden and a settler's nightmare. Ruthlessly exploited by the Atlantic world, it had also been taken into that world's treasury of metaphor and knowledge. A growing science would shape both the exploitation and the preservation of the land under its new masters.

4

The Row-Crop Empire

THE EXPANSION OF THE SOUTH ACROSS THE APPALACHIANS AND THE MISSISsippi River to the fringes of the high plains was one of the great American folk wanderings. Motivated by the longing for fresh and cheap land, and by obscurer urges, such as simple restlessness and the large human capacity for dissatisfaction, southerners completed their occupation of a region as large as western Europe. Despite the variety of the land—which contained regions of pine barrens and prairies, of hardwood forests and limey plateaus, of some of the world's oldest mountains and a considerable part of its third longest river—and the variety of the societies from which they came, the settlers of the Southwest had certain broad similarities. They might be farmers large or small, but most farmed or lived by serving the needs of farmers. Their way of dealing with the wild assumed each man's right to use beasts and timber as he saw fit. Not all owned or ever would own slaves, but most accepted slavery as a mode of holding and creating wealth. Throughout the Southwest a burgeoning democracy amplified the folk voice for both good and ill.

Southwesterners were a mixed population, drawn from the most varied sources, finding occupations as primitive drovers and city dwellers, yeoman farmers and vavasours of great plantations. Together they formed a growing mass, which between 1820 and 1850 displaced the center of southern population and political influence from the eastern seaboard. As the South in terms of population grew smaller within the nation, the Southwest grew great within the region. Vigorous and sometimes heedless, the new region expanded to continental scope the rowcrop empire which the Tidewater had founded, and transformed it with western democracy and western energy. The impact on soil, forests, rivers, and wildlife was great, in the sense that much land was cleared and the seed of some intractable problems drilled into it. Yet immediate changes were limited by low population density and the disorderly looseness of the new society.

Transappalachia was a varied and inviting region whose invasion by settlers followed quickly on their movement into the Piedmont. The rise in population in the colonial Chesapeake during the early eighteenth century brought migration to western Maryland, the valleys of Virginia, and the Southside. In large measure the Great Valley was settled by immigrants moving southward from the middle colonies, who sometimes spilled over the Blue Ridge to settle the Piedmont from the west. Scattered settlements had appeared in backcountry Maryland and the Shenandoah during the 1720s, and the alternations of war and peace during the century that followed made no lasting impression on the flow to the west and south. Between the Seven Years War and the Revolution settlers penetrated the mountains, often ascending streams whose headwaters brought them within striking distance of the Monongahela and the Ohio. The penetration of Kentucky and Tennessee went on during the 1770s despite Revolution, Indian resistance, and intrigues by the British at Detroit. Harrodsburg and Boonesborough were established in 1775, and helped to screen Nashville (established in 1779) against attack from the north. Independence found the transappalachian southern frontier a reality.[1]

The first permanent settlements in the South Carolina Piedmont were made in 1740–1760, and in the decades that followed, streams of settlers flowed onto the uplands. Repeated emigrations, land sales, and replacement of pioneers by larger planters made a complex pattern, exactly the same in no two places. Some groups leapfrogged ahead over hundreds of miles of wilderness, like the immigrants from South Carolina and Georgia who began to settle the Natchez region on the east bank of the Mississippi after 1765. Settlement of the inland Gulf South was halted by extensive Indian domains, and awaited the signing of treaties that displaced the original owners westward. About 1790 Oglethorpe County, Georgia, was the western frontier, "a land of scattering log cabins, range livestock, horse and cattle thieves, and Indian depredations." Cessions by the Creeks in 1802 and 1804 moved the line from the Oconee to the Ocmulgee River, then halted for almost two decades until the Creek nation yielded again.[2]

For evident reasons, newly settled areas began as regions of subsistence farming upon which commercial farming was superimposed as means of transporting crops (or of making crops transportable, as by turning corn into whiskey) were devised. In 1784, John Filson recorded the existence at eight settlements in Kentucky, and "many other places" not named, of "inspecting-houses . . . for Tobacco, which may be cultivated to great advantage; although not altogether the staple commodity of the country."[3] The entry of the staples was gradual, the farms that produced them small. Corn was the pioneer, here as elsewhere "more plausibly the symbol of the United States than . . . the bald eagle." Early arrivals in Transappalachia entered the skin trade and hunted the buffalo and deer, exterminating the first and depleting the second. For a time pelts played the role of money, as tobacco had in the early Chesapeake; in the "state of Franklin" in present-day Tennessee, salaries of the governor and the chief justice during 1784 were fixed in deerskins. Commercial agriculture developed slowly, but by the 1790s slaveowners were entering the region, farmers were seeking markets for their surplus, and a money economy was replacing barter. Cutting of timber went apace, and by the 1820s, in the region around Lexington, Kentucky, the need to conserve was being discussed.[4]

The early primitive settlements changed quickly. Viewing the site of Louisville in 1796–1797, Moses Austin exclaimed with dismay that "the handy work of Man has insted of improving destroy'd the works of Nature and made it a detestable place." Yet he saw a bustling future for Kentucky, "for it is Not possible a Country which has within itself everything to make its settlers Rich and Happy can remain Unnotic'd by the American people." Progress was swift in the early nineteenth century, as cabins inhabited by men in hunting shirts gave way to houses in the towns and plantations beyond them. In general, the Great Valley became characterized by a solid, yeoman prosperity; the mountains proper and the Cumberland Plateau, by enduring poverty and isolation; the Bluegrass region by hemp and grazing; while the Cotton Kingdom was quick to claim the southern part of the Nashville basin and southwest Tennessee. The great tree of the Mississippi River system, in particular its branches the Ohio, Cumberland, and Tennessee, bound accessible parts of the region into worldwide commerce. Beyond the mountains the Atlantic world had gained immense new provinces.[5]

In 1803 the Louisiana Purchase added to the South a new city, a new culture, and a new physiographic region. Indeed, several regions, for beyond the western escarpment of the alluvial valley stretched forested hills giving way to treeless plains which were to mark for the white and black South, as they had for the red, an indefinite yet enduring cultural boundary.

Dominating the immediate area of settlement was the Mississippi, which had created the alluvial valley. From Missouri's Commerce Hills to the sea 600 miles to the south, the river expanded during floodtime to cover vast areas of bottomland, precipitating its heavier alluvium to form natural levees, and carrying finely-divided clays back into the swamps. Human settlement gravitated to these natural levees to be safe from floods, to exploit their fertility, and to use the river for transport. The Franco-Spanish Creole culture of the valley showed in this respect certain similarities to the Mississippian Indian culture, whose remnants it encountered among the Natchez and helped to destroy.

The early history of French Louisiana had shown marked similarities to that of the English settlements along the Atlantic coast. Settlers searched for a staple, introduced forced labor, and experienced disease. The soil of the first settlement around Biloxi on the Gulf Coast was infertile, the whitish stuff of the coastal barrens. Mosquitoes were abundant and malaria was quickly introduced, an official noting in August 1701 that he had thirty sick who could not "recover from a tertian fever that saps their strength for lack of remedies and food." Typhoid was a probable early colonist, as well. While the settlement sickened and searched for mineral wealth, the early development of the alluvial valley was largely in the hands of traders and herdsmen—not only Frenchmen, but Englishmen owing allegiance to Charles Town.[6]

Agriculture developed in the next century. In 1718 the French transferred their colony from "the sterile lands of *Biloxi, Mobile,* and *St. Louis Bays,* to the rich country bordering the Mississippi." An important herding industry grew up in southern Louisiana as it had earlier in the Illinois country. Native species continued to be exploited for food and skins, as hunters in the St. Francis River valley and elsewhere sent buffalo tongues, saltmeat, and bear oil downriver to New Orleans. In 1762 the town exported bear oil and tallow to the value of 25,000 livres. Agriculture—leading quickly to the discovery of staples and the widespread introduction of slavery—became the main business of lower Louisiana under the guidance of the Compagnie de l'Occident. The moving of the capital had been the first step; the introduction of a mixed bag of settlers was another, as free colonists, indentured servants, felons, exiles, and slaves were dispatched to the Gulf. By 1721 black slaves formed almost half the population of New Orleans; a few years later they were more than half.[7]

The settlers grew rice because the land behind the levees was easily flooded, and maize, which as usual established itself as the food of the poor. A modest tobacco industry grew up, and indigo, as well, until destroyed in the 1790s by Asian competition and local insects. Experiments began with sugar cane, a much-travelled plant, which had been carried from India to Spain by the Muslims, by the Spanish to the New World, and by the French from Haiti to Louisiana.[8] In the growth of this commercial agriculture, the Mississippi played on a greater scale the role of the Chesapeake waterways, extending the Atlantic world, its commercial demands, tastes, and fashions, deep into the interior of North America. French, English, and later American towns up and down the river system accumulated the products of the forests and the soil for shipment to New Orleans.

Settlers early found that the river was not altogether a blessing. Its spring rises could top the natural levees, and the response of the colonists was to raise them higher by artificial embankments. In laying out New

Orleans, Sieur Blond de la Tour "raised in front . . . a dike or levee of earth." Under a rule that riparian landowners must build and maintain levees to retain title to their property, the levees spread up and downriver from the settlement. Here was a development that tended to fix the course of the river. But in other places the settlers wished rather to change it for easier navigation. This was a tempting notion because the river so often changed its own channel, cutting off meander loops in time of flood. Charlevoix noted two sites on the lower river called Pointe Coupee; the second took its name from an early artificial cutoff when "Canadians, by means of digging the channel of a small brook . . . carried the waters of the river into it, where such is the impetuosity of the stream, that the point has been entirely cut through, and thereby the travellers save fourteen leagues of their voyage."[9] Over the centuries the determination of those who wished to build permanent settlements beside the river to fix its course, restrain its floods, and improve its navigation became the motive for widespread and enduring environmental change.

The Louisianians' experience of disease resembled that of the English colonies. Climate, landform, and the invitation to trade which the river presented all encouraged diseases both endemic and epidemic. The mixing of populations went extraordinarily far, even by North American standards, as Indian, French, Spanish, Negro, Anglo-American, German, and many smaller immigrant strains, with their domestic animals, mingled in a region well suited to the transmission of a wide variety of ailments.

Animals no less than men showed the effects. A striking difference between the Indian agriculture of Louisiana and that of the east was the abundance of livestock. Wild horses had followed the traditional route of Meso-American travellers north from Mexico, and in consequence tribes west of the Mississippi had many horses. In the seventeenth century Henri de Tonti had seen horses near Nachitoches still marked with Spanish brands. Pigs presumably left by De Soto had multiplied. Antoine Simon Le Page du Pratz reported that all the animals brought from France "have multiplied and thriven perfectly well." At the same time, native animals swarmed in great numbers; in the eighteenth century, large herds of buffalo still browsed the alluvial valley, forming "the chief food of the natives, and the French also for a long time." The mixing of wild and domestic, Spanish, French, and Indian livestock may have played a role in the 1748 epizootic which killed so many cattle that a shortage of beef resulted.[10]

Human beings suffered severely from ailments of their own. A heavy mortality among newcomers continued until the twentieth century; Indians as well died rapidly. Again there seems good reason to indict the immigrant ships, which were, as usual, excellent incubators of disease. Mortality in the slave ships resembled that described by Garden at Charles Town. Yaws and scurvy were common among newly arrived slaves, and yellow fever, common in the West Indies, may have been introduced as early as 1704. A severe epidemic occurred in 1796, marking the growth of New Orleans to a size large enough to support more than the random cases which may have occurred earlier. Counting only from this outbreak, the disease was to have a reign of more than a century in lower Louisiana, with occasional forays into the river ports above. It remained an import, its outbreaks episodic, but the episodes were frequent and terrible, and its reputation even more than its actual presence lay upon the region like an incubus. Having lost his private secretary, many friends, his wife, and his daughter to the epidemic of 1804, Governor William C.C. Claiborne wrote his own judgment of the area: "Lower Louisiana is a beautiful Country, and rewards abundantly the Labour of man;—But the Climate is a wretched one, and destructive to human life."[11]

Overall, early mortality in Louisiana rivalled that at Jamestown, and the effects of disease on the Indians likewise resembled the experience of the Atlantic coast. Writing in 1725, Bienville remarked that Louisiana "was formerly most densely populated with Indians but at present of these prodigious quantities of different nations one sees ony pitiful remnants" by reason of intertribal war and "diseases that the Europeans have brought into the country and which were formerly unknown to the Indians." Le Page du Pratz found that smallpox and colds (influenza?) made "dreadful ravages" among the Indians. The Natchez, wrote another Frenchman, had diminished by a third during their first six years of contact with the French, "so true is it that . . . God wishes to make them give place to others."

As usually, the Indians had cleared and appropriated much of the best land in the valley, which upon their death became available to whites without the trouble of conquest. In the early nineteenth century an American traveller noted in what had become the Arkansas Territory "considerable tracts of good and elevated land, once numerously peopled by the natives" and now deserted. War with the French was a factor in the decline of the tribes; the Natchez, a remarkable people preserving much of the Mississippian tradition, were finally destroyed and the survivors dispersed by French soldiery. But under the circumstances the impact of disease could hardly have failed to be great, and much testimony suggests that in the alluvial valley the tragedy of the Atlantic coastal tribes was repeated.[12]

Yet where so much resembled the Atlantic coast, the river was a dominating environmental fact that helped to make the French dominion different in important respects from the English. It encouraged population to disperse more widely, while the Appalachian ridges had tended to constrict the coastal settlements. The river also gave New Orleans and other river towns throughout their early history an excellent source of potable water. Though Mississippi water met none of the preferences of the eighteenth century, being muddy in an age when clarity was supposed to imply purity, the lack of heavy upstream settlement and the dilution of impurities by the sheer size of the stream made it uncommonly healthful, a fact often noted by contemporaries, though not understood. In this, at least, the region was fortunate by comparison with the pioneer settlements of the eastern seaboard.

Other characteristics of the valley, soon to be discovered by settlers pushing into eastern Tennessee and the Mississippi Territory as well as those in the Louisiana cession, were its susceptibility to storms and earthquakes. Hurricanes were early visitors to New Orleans. Movements in the faulted bedrock of the valley also made for some notable oscillations in the landscape. The earthquake of 1811–1812, centered near New Madrid, Missouri, became one of the celebrated calamities of the early nineteenth century, destroying relatively few lives only because there were so few inhabitants in the area. Tremors were felt through much of the eastern United States, though the central river valleys were the worst affected; a Louisville engineer counted 1,874 shocks between September 1811 and March 1812. Besides physical changes which included a temporary upstream flow in the Mississippi and the gradual disappearance of New Madrid into the river, a variety of social phenomena from lawsuits to religious conversions were associated with the tremors. Memberships in the Western Conference of the Methodist Church, for example, are said to have risen 50 percent in the hardest hit region, while the number of communicants in the rest of the nation increased by less than 1 percent.

Overall, the purchase and occupation of Louisiana brought into the South a national highway, a province of great potential wealth, and a land singularly afflicted by many natural ills. In all three aspects it was to modify greatly the character and future of the region.[13]

Across much of the new territory south and east of the Appalachians rose the phenomenon of the Cotton Kingdom. The development in the late eighteenth century of an efficient gin for separating the lint and seeds of short-staple cotton is a justly famous example of the impact of technology on culture. It is also true that the culture was in search of the technology. To fuel an expansion of the character and speed that occurred between 1790 and 1837, the South needed some commercial crop adapted to the climate, demanded by the overseas market, and suitable for production in circumstances ranging from the frontier farm to the great plantation. The fact that the English textile industry was already mechanizing made this crop particularly timely. Cotton as a great staple was invented as much as grown, made to order for the place and time. Early in the nineteenth century Mexican cotton was introduced, because its "large wide-open bolls" facilitated picking and greatly increased the amount that a worker could gather in a day.[14]

Like the crop itself, the boundaries of the Cotton Kingdom were determined by tacit agreement between man and nature. Antebellum opinion sometimes held that, roughly speaking, the southern border of Pennsylvania, extended west, marked the limit of profitable production. With few exceptions, however, the actual dimensions of the region were smaller. A better boundary was the 77°F summer isotherm. This line runs roughly from the northeastern border of North Carolina, dipping south of the Appalachian massif, and rising to the northwestern border of Tennessee. West of the Mississippi it skirts the Ozark highlands and rises again to north-central Oklahoma. On occasion cotton was grown commercially north of this line in response to high prices—notably in southern Virginia and the Nashville basin. Noncommercial production for household use was also common. But in general terms, this was the boundary of the Cotton Kingdom. Northward, cotton grew well enough in mild years but poorly in cool ones; in short, it was not a money crop that could be depended upon. The line was a limit defined partly by the human need for consistency in a commercial undertaking, partly by the nature of the plant itself.[15]

The dynamism of such a boundary is evident. The weather varied, the market varied, techniques of farming varied, and in consequence the line exhibited only an average stability. Nevertheless, it was sufficient to insure that cotton became the predominant cash crop of the Carolina Piedmont and much of the great Southwest, and this in turn stamped a unique character upon the life of the region. Both for household manufactures and as a source of cash, both as the poor man's source of pocket money and as the plantation owner's way to wealth, the plant was eagerly adopted. In 1801 Governor Claiborne estimated the value of the crop in the Mississippi Territory at $700,000, suggesting production of about three million pounds. Though interrupted by wars, embargoes, depressions, and market variations, the nation's production of raw cotton climbed from 3,135 bales in 1790 to 208,986 in 1815. It reached 4.5 million bales in 1861.[16]

Possessing in this staple a most convenient way to create a new South modeled on the old, settlers began again to push back the frontier of the Gulf region. Though opened to settlement in 1798 with the creation of the Mississippi Territory, the Old Southwest, with its strong Indian tribes, and with the Spanish still controlling Florida, at first developed slowly. Defeat of the Creeks, the end of the War of 1812, a series of Indian removal treaties, and the postwar boom in cotton prices brought a rush of settlement. An observer on the Piedmont noted "immigrants—if there be such a word—farmers errant, proceeding with all their worldly goods, according to the usual tide of these matters in this country, from East to West, or rather, to be quite correct, from North-east to South-west—from Virginia and Maryland to Florida, Georgia, and Alabama."[17]

Behind the farmers-errant came plantation owners and their slaves, the whole human mass beginning to shift the balance of population within the South from east to west. Whether driven by soil depletion, by an over-rigid and structured social order that had engrossed the best lands, or by simple restlessness, the *Völkerwänderung* was in regional terms a huge one. Between 1800 and 1860 Georgia's white population increased about nine times, Louisiana's and Tennessee's about ten times, Mississippi's about one hundred times, Arkansas's over four hundred times, and Alabama's about one thousand times. (Such figures of course indicate not only the rapidity of settlement but the initial lack of a white populace in the middle Gulf region.)

Meanwhile the population of Maryland, Virginia, and the Carolinas also increased, but much less than the national average. The 1830s were North Carolina's decade of heaviest emigration. Thirty-two of the state's sixty-eight counties lost population; as a whole the state increased by only 2.5 percent, though it had one of the nation's highest birth rates. Many of the emigrants moved north of the Ohio River, a fundamental motive being, as one of them put it, to exchange "them old wourn out red fields" for the "perraria" of Illinois where the soil "is as rich and black as you wold want it." By 1850, four of ten living South Carolinians were emigrants from their native state, as were about a third of the North Carolinians and a quarter of the Georgians.[18]

The Cotton Kingdom made rapid conquests, not only in the middle Gulf region, but along the lower Mississippi River and its tributaries. Prior to the War of 1812 cotton production in Louisiana had been comparatively small (two million pounds estimated for 1811). But the fiber was already being grown in the Natchez area, and between 1810 and 1820 it made rapid progress in Louisiana, as well, spreading into the Attakapas and Opelousas regions and up the Red River. Soon settlers, led by squatters and harassed by federal troops trying to protect Indian lands, were entering the region where Louisiana, Arkansas, and Texas later met.

Frontier life in Arkansas now reproduced many features of Kentucky and Tennessee in the late eighteenth century. Early settlers hunted, trapped, and grazed cattle; the Indian presence was still considerable, and farmers were few. But close to the rivers scattered settlements sprang up on areas of superior land. Before 1820 "Curran's settlement" on the Arkansas River was producing 1,000–1,500 pounds of cotton to an acre, "of a staple in no way inferior to that of Red River." Otherwise the life was typically that of a southern frontier; cattle roamed wild, feeding themselves overwinter on cane and rushes, and in slaughtering time were "killed just as they are hunted up from the prairie or the cane-brake." By the Civil War the eastern and southern alluvial lands were part of the Cotton Kingdom, and the staple was also grown by small farmers in the uplands; stockraising continued, and wheat and orchards were common.[19]

Much of Louisiana, however, profitably continued the growth of sugarcane, whose introduction dated from the French, and successful large-scale production from the Spanish dominion. The basic soils of the sugar country were far less varied than those of the Cotton Kingdom. Essentially cane grew on recent alluvium in the lower Mississippi and Red River valleys and on the Pleistocene terraces above them. Since cane depletes potassium rapidly, the terraces required frequent fertilizing, while the lowlands, though sometimes ill-drained, were too rich to respond to most fertilizers even if applied. Sugar growing evolved from its colonial beginnings as a remarkably artificial endeavor. Sustained by tariffs in a region too far north to be truly competitive with the Indies, and dependent on machinery and expertise of a high order in the refining process, sugar production was as much an industry as a form of agriculture.[20] Here, as in the rice lands of the Carolinas and Georgia, a highly rationalized endeavor employing great numbers of slaves on large units existed on the borders of the more rough-and-ready Cotton Kingdom.

To the west and south of the region lay borderlands where climate and history delayed the march of settlement. Under the Spanish empire, Florida had not developed extensively, except as a refuge for Indians, slaves, and whites who had their own good reasons for leaving the English-dominated regions to the north. Twenty years of English ownership (1763–1783) brought in some planters and herdsmen, but the retrocession returned this oldest settled area of the South to the status of an imperial buffer zone. After many alarums and Jacksonian excursions, the United States acquired the province in 1819, and a "considerable immigration of Americans" followed. Planting of cotton and tobacco expanded in the north, and a herding industry took advantage of the unoccupied lands. Much of the best land in the territory, however, belonged to the Seminoles, a people formed from refugees of broken northern tribes. Probably feeling that they had been driven far enough, the Seminoles put up a protracted guerrilla resistance to removal in the 1830s and early 1840s, and again in 1855–1858. Their resistance was favored by desperation, topography, and disease, especially malaria, which caused a soldier obliged to fight them to pronounce the future Vacationland "a most hideous region."[21] Florida was a slow developer among the Gulf states, and only the northern border became part of the Cotton Kingdom.

The penetration of Texas by American settlers began even before 1821, when Moses Austin entered a contract with the Spanish for settlement along the Brazos River which his son Stephen carried out a few years later. Determined efforts by Americans preserved slavery in the region and made the development of a plantation system possible. Yet herding and small farming were the occupations of most Texans until after the revolution of 1836. Then planters and their slaves settled more thickly, most in the alluvial lands along the small rivers. The fame of the unused soil and the speculative fever that developed after 1840 brought rapid immigration; in the 1850s total population almost tripled and the slave population multiplied three times over. By and large, however, the settlers still had not broken the best lands of the state—the grass-covered soils of the prairie. The river basins of the Gulf plain, and the Red River country, both linked to New Orleans by steamboat, together with regions in the oak uplands of the coastal plain, formed the antebellum Texas cotton bowl.

So the new—and, as it turned out, geographically the final—South took form. Overall, the region was a land statistically dominated by small farmers; its principal crop was corn; after the 1830s its favored politics, outside the tenacious squirearchies of Virginia and South Carolina, were largely popular and democratic, though remnants of an older conservative order lingered also in Kentucky, Maryland, and North Carolina. Wealth followed the great staples, and slave ownership, with few exceptions, marked the successful and influential man. Despite the rise of political democracy, the basic economy and social system of the southern seaboard colonies had been transplanted westward into a realm of almost continental scope. Among the staples, the *arriviste* cotton had become dominant, supporting a thriving class of new men who infused regional life with their intense local loyalties and the animus of their populistic conservatism. The endless fields of the row-crop empire and the attitudes of its masters were both to make a heavy and lasting impact upon the southern environment.

The treatment of soil in the empire is not easy, or perhaps even possible, to summarize. Once portrayed as the haunt of a tedious monoculture, the antebellum South displayed impressive variety in its modes of farming. It was not only crops that varied in the South, but the kind of agriculture, exhibiting a deep though by no means simple or deterministic relationship between society, climate, and landform.

In the Maryland Piedmont and in parts of Kentucky and northern Virginia, a system of mixed farming and stockraising developed which extended the methods of Pennsylvania into the South. Such border and Great Valley regions, with their relatively cool climates, low slave populations, dairy cattle, stone barns, and fat silos were essentially outside the dominant southern system. A region of intensive agriculture was the area in South Carolina and Georgia, all within thirty miles of the ocean, where sea-island cotton was grown. Yields of the glossy long-staple product, a high-priced luxury fiber, were maintained by fertilizing the fields with marsh mud and other soil-restoring materials. Ricelands both along the Atlantic coast and in Louisiana were restored by flooding, though the gifts of the rivers were not always desirable, especially when upstream deforestation and overcropping exposed the sterile layers of the subsoil to erosion.[22]

Even in the worst depleted regions of coastal Virginia and Maryland, farming changed and improved through the use of fertilizers both mineral (plaster, marl) and green (clover, cowpeas). Much of Tidewater Virginia shifted from tobacco growing to wheat, and in Maryland the weed was reduced from the dominant staple to one part of a system of mixed farming. Eastern Maryland, with its urban focus in Baltimore, its fishing and manufacturing, had evolved since the eighteenth century into a complex society with close resemblances, despite slavery, to the North. The countryside responded to the influence of the growing towns, finding an internal market whose priorities wee quite different from those of the Atlantic trade.

Far more "southern" in the usual sense was the farming of the Atlantic Piedmont. As defined by geographers, the Piedmont runs from mid-Virginia to eastern Alabama, a sloping region drained by east- and south-flowing rivers. With elevations ranging from 200 to 1,500 feet and heavy rainfall (especially at its southwestern end), the province is exceptionally liable to erosion, the more so because of its friable, deeply weathered soils. As the wave of settlement passed south and west, and especially as cotton planting destroyed natural groundcover without replacing it, erosion increased. By conventional measures of erosive intensity, row-crops grown without conservation measures permitted 500 times that of deep forest cover. Erosion in the region spiked suddenly between 1800 and 1860, with the eastern South Carolina upland apparently the area hardest hit, from a combination of early settlement, poor farming practices, susceptible soil, and high rainfall.[23]

The pattern was clearly the consequence of conventional southern frontier practice imposed upon a sloping and fragile region. Farmers repeatedly cleared new fields and abandoned the old, permanently or to an extended fallow, in the familiar pattern of primitive exploitative agriculture. Cattle turned loose to graze on land supposedly "resting" then did further damage, for much of the erosion surely was over fallow, not over the working fields. Stanley W. Trimble's recent study of the erosive process in the Piedmont has indicated, however, that the maximum effect came during the two generations that followed the Civil War. The antebellum period saw the transition from very low natural levels of erosion to the very high ones of the Cotton Kingdom which would eventually strip the region of an estimated six cubic miles of topsoil.

Throughout the South complaints of planters about soil exhaustion were commonly heard, at least until the late 1840s, when sectional antagonisms had risen so high that southerners hesitated to criticize any aspect of their section or society. Voices were raised even in naturally rich and broadly atypical regions like the Great Valley.[24] But the peculiar difficulties of southern agriculture are not easy to fix. Cotton is not a particularly voracious plant; it uses, for example, less phosphorus and potassium than alfalfa. If a crop is sought which worsened the natural problems of the soil, surely it was the South's most abundant crop, corn, that should be impeached. Corn takes thirteen nutrient elements from the soil and is a particularly heavy user of nitrogen and phosphorus. Two-thirds to three-fourths of the plant's uptake of these elements enters the grain as it matures and so is removed by the harvest. Both as a row crop and as a direct destroyer of fertility, the South's basic food, more than its commercial staple, injured the soil.

Another difficulty lay in the effect of heat on cattle. Julius Rubin has suggested that the fundamental divergence between southern and northern agriculture lay in the relative absence, especially in the lower South, of the mixed farming and stock-raising enterprises of many northern regions. Instead, an extensive, constantly shifting system focused on the staples, and a range-cattle industry exploiting forests and abandoned fields grew up. Such a regression to the simplest forms of the past was to be expected in the first years of settlement. That the system endured so widely and so long was at least partly due to the weather. The heat of the lower South limited milk production in the breeds of cattle which were then available. The quick-growing wild greenery was high in fiber content that bloated animals without nourishing them. Lack of dairying and the production of high-grade meat as an integral part of agriculture eliminated the contribution that both herds and fodder crops might have made to the soil. This, in turn, helped to compel heavy purchases of commercial fertilizer for those who could afford it—a factor in the guano mania of the 1840s and 1850s. More commonly, soil depletion reinforced the shifting agriculture already described, implying a low population density as well as maximum impact on forests and forest ecologies. Low population, static rather than growing towns, lack of urban markets, and the difficulties and costs of transport in a half-settled region in turn reacted upon the agricultural base which had helped to create them.[25]

Yet before the end of the eighteenth century, English methods of agricultural reform had begun to attract attention in the South. Washington, Jefferson, and John Taylor continued to espouse reform, though with mixed success, even on their own land. The next century saw the spread of agricultural societies and the publication of many farming journals. The principles of Sir Humphrey Davy's agricultural chemistry filtered via the work of reformer-publicists into the consciousness of producers. The problems of erosion, especially in the Piedmont, were much discussed, and contour plowing made its way "against a storm of ridicule" to find widespread acceptance in uplands from Maryland to Texas by the time of the Civil War. Agriculture reform was effectively combined with sectional patriotism, notably by Edmund Ruffin, the prophet of marl as an antidote for the region's acid soils. Speeches at agricultural fairs deplored erosion, demanded reform, and emphasized the role of science in restoring state and section to a presumed earlier condition of prosperity. As for private motives, the aim of much reform and especially of crop diversification was to limit expenditures by making the plantation self-sufficient. Though such movements were at first associated with the depleted eastern soils, a wave of reform spread during the late 1830s and 1840s into the Southwest, along with interest and investment in fertilizers.[26]

The best of these, and the most expensive, was Peruvian guano. For the large planter who could afford it, this and other soil additives became a feature of rationalized agriculture practiced by gang labor. Fertilizers were and remained of special importance to southern soils, most of which lack nitrogen and potassium and are slightly to strongly acid. Fertilizing, however, could easily take on the character of a technological "fix," especially when so little was known of the chemistry of soils, and depenence on guano could make good husbandry and erosion control seem less urgent than they actually were.[27]

Like any development adding new costs to agriculture, fertilizers also favored some farmers over others. Difficulties in transport and high costs insured that through great areas, especially of the inland South, guano would not be used, at any rate not in the quantities required. Costs favored the astute planter with wealth and slaves who could achieve commercial success by gang labor, economies of scale, and heavy fertilizing. Meanwhile, because the whole process of farming was labor-intensive, the growth of a rural population so largely made up of slaves, whose function it was to work but not consume, limited the development of the region as a market for goods and helped to prepare the paradox of the postwar rural South. In a region generally deficient in people, too many were living upon the land, often in direst poverty, for such was the invariable condition of the slave. This was productive efficiency at a price, and part of the price must have been paid by the shrinking areas of wild land, where hunting and fishing helped supply the tables of the master, the overseer, and to some extent the slave.

As long as prices held and erosion was a process only beginning, the Cotton Kingdom throve. Acreage, production, and yield all grew, and would continue to do so until 1890 or so. Especially during the 1850s the mood of the cotton planters was ebullient, as reflected in their journals. Profits were good and "slavery itself was generally returning high profits to those who invested in that peculiar institution." The question of whether the antebellum South was able, in the face of its commercial commitments, to feed itself remains in doubt. A complex of many subregions, the South defies summary on this point; by and large it probably did so, though considerable food was imported from the West, as well.[28]

The limitations of the Old South in political, social, and economic terms need no repetition. The habit of expansion begotten of the experiences of the region's first quarter-millenium encountered a solid obstacle in western territories which northern farmers were determined to possess for themselves and for people like themselves. Both North and South were dynamic, but so greatly did the pace of change differ that the latter seemed to be standing still. The differences between them grew more notable than ever during the very height of the row-crop empire. Almost 80 percent of the southern labor force was in agriculture, as against 30 percent in the Northeast and 54 percent in the Midwest. The service sector was half that of the Midwest, and the manufacturing sector less than half. "Indeed," writes Harold D. Woodman, "the economic structure of the South in this respect very closely resembled the structure of the poorest of present-day undeveloped nations." In the mid-nineteenth century world this made the region more representative of the general human condition, not less. But the South was linked to the explosively developing North, with its natural wealth, its flexible labor system, its cool climate, and its consequent influx of immigrants.[29]

The expansion of the South had brought the region much, but it may be worthwhile to note what, in terms of resources, it did not bring. In the Great Valley, the river valleys, the Bluegrass region, and the black belt, southerners had found fine soils, but in the other elements of a vigorous modern economy southern pioneers had been less fortunate. Southern rivers were useful for commerce but most were of little value for power production with the technology of the times. The South's true mineral wealth—coal, Alabama hematite, western petroleum, and the like—would wait one or two generations for exploiting, and in some cases even for discovery. No equivalent had been found to the deposits of lead, copper, and zinc announced in 1839 by David Dale Owen on public lands in Wisconsin and Illinois. Even the enclaves of rich soil were enclosed in the usual sandy or hilly regions and pine barrens of poor to mediocre land. These were fundamental limiting factors.[30]

Triumphant agriculture took out heavy liens against the natural dower. Row crops bared the soil, the rows made watercourses for the rains, which were heavy, and the colonial practice of plowing straight up and down hills was by no means extirpated. Further, any system which covers too many fields with the same plant falls afoul of the ecological principle which states that the simplest systems are apt to be the most unstable. Natural systems, almost always complex, contain multitudes of checks which prevent any single event from threatening the whole with destruction. In any great center of monoculture, soil toxins develop and parasites of many sorts are encouraged to multiply explosively. The South was not unique in planting great areas to a few basic plants, but no more than the Ireland of the 1840s was it exempt from the dangers inherent in such dependence.

Few can view without sympathy the extraordinary achievements of the black and white pioneers and settlers who created the row-crop empire in a few generations of prodigious effort. To make a living from the land without injuring it is nowhere easy, and in much of the South probably harder than elsewhere in America. It does appear, however, that the westward movement had imposed a burden on the southern environment which neither wisdom nor good will would be able to lighten significantly for generations to come.

5

Exploitation Limited

THE CONQUEST OF THE SOUTHWEST BY WHITE SETTLERS AND THEIR BLACK chattels impacted on every aspect of the environment. Diseases carried into the new regions took their familiar course with Indians and settlers alike. Hunters fell upon wild species with their customary vigor. Herds of cattle grazed the prairies and the open pine woods. Forests were cut, often merely to get them out of the way; the beginnings, though only the beginnings, of a commercial lumbering industry also appeared. Every part of the primeval landscape—woods, soil, and waterways, and the animals that used them—was altered by the demands of a society with many wants of its own, and foreign markets to serve, as well.

Yet a relatively low population and embryonic industrialization limited the changes that resulted. So did the South's evolving ideology of localism, with its jealousy of power wherever centralized. Laissez-faire was, however, at best an ambiguous friend to the wilderness, and a positive enemy of orderly development. Failure to use any common property to the full redounded, not to the common good, but to the profit of other less scrupulous exploiters. The sketchy colonial attempts to regulate hunting apparently were pursued with less vigor. The South tended to exalt, sometimes with a special anarchic heedlessness, the contemporary American standard of exploitation without limit. Yet limits did exist—the physical restraints imposed by distance and disease, and the social ones of persistent ruralism, disorganization, and incompetence.

Disease spread with settlement into the new regions. One that had long seemed endemic—yaws—proved to have been an exotic after all, for it disappeared about the turn of the century, coincidentally with the end of the legal slave trade. But malaria increased in the Piedmont, the river valleys, and the Gulf coastal plain, where areas of infection already existed around the French and Spanish settlements. Among the settlers of the central river valleys it was known under an amazing variety of names, which Dr. Daniel Drake of Cincinnati cataloged—"autumnal, bilious, intermittent, remittent, congestive, miasmatic, malarial, marsh, malignant, chill-fever, ague, fever and ague, dumb ague, and lastly, *the* Fever." However named, it was, Drake noted, "the *great* cause of mortality, or infirmity of constitution, especially in the southern portions of the Valley."[1]

Mixed with malaria no doubt was typhoid. In 1807 such an outbreak occurred in May's-Lick, Kentucky, as a "fever of the typhous or typhoid kind" attacked "every house in the village, and . . . many in its vicinity." Moving with the march of settlement, and favoring the rivers, the febrile diseases by 1819, if no earlier, had reached the western edges of the settled region. Thomas Nuttall recorded that "the ague and bilious fever spread throughout the [Arkansas] territory" between July and October; the naturalist himself was taken violently ill of a gastrointestinal complaint, while the post surgeon at Fort Smith died of a "nervous fever" and a visiting missionary of a "lingering fever." Hardest hit were the Cherokees, 100 of those who had been removed to the banks of the Arkansas dying of the fever which, like the whites, had followed them.[2]

The western settlements were now experiencing what the coastal towns and hamlets had gone through decades or centuries before. Commonly an early record of good health would be followed by unexpected outbreaks, after which the fevers and gastrointestinal complaints would settle down to a pattern of seasonal endemicity. A potent source of disease must have been the extraordinary mixing of peoples that characterized the newly settled regions. Even on the Atlantic Piedmont a mixture of English, Scotch-Irish, German, and African immigrants was commonplace, and American states from Pennsylvania to Georgia were represented. More complex was the situation of the Mississippi and Ohio valleys, with their direct water connections to the sea. Drake remarked that "the world has not before witnessed such a commingling of races"—comprehending in that term race, culture, and national origin—which he found "a living compound, as yet in the forming stage." Nuttall too spoke of the "jarring vortex of heterogeneous population" that was shaping the western country. The great river and its tributaries linked many small settlements into a single disease community, and the epidemics of the port cities, like their goods, were transmitted to tiny river landings, remote and seemingly safe towns. In Warrington on the Mississippi south of the Arkansas River, a hamlet composed of "two inns and as many stores," thirty-seven people died of yellow fever in 1819, the disease "said to have been introduced by the steam-boat Alabama."[3]

The invasion of disease knew neither North nor South; Indiana, Ohio, and Illinois were harried, as were Kentucky, Tennessee, and the Gulf states. Climate and topography, however, soon began to make themselves felt. Natural processes marked off zones in which certain ills were well established, others where their existence was marginal and precarious. Especially was this true in regard to malaria. So complex is the life-cycle of plasmodium, so dependent on probabilities—the number of mosquitoes, the number of their contacts with infected persons, the number of days when the temperature is right for both parasite and vector to multiply—that the survival of the disease outside of very warm regions is statistically dubious at best. The evolution of frontier settlements toward greater order and cleanliness, improvements in drainage, and the advent of tighter houses conspired against it. So did northern agricultural patterns, whether in the North or in the atypical portions of the South where they also took hold. Wherever cattle were penned close to human dwellings, anophelines could find an alternative blood source which they preferred.

Science as well as agriculture worked against the plasmodium. In 1820 quinine sulfate was isolated by French chemists. Effective dosage was long in doubt, but in the 1830s southern physicians tried—and the army, during the Seminole Wars, confirmed—that large doses could prevent the paroxysms of malaria. As "tonics" made with quinine became popular in the United States, adding to the strictly medicinal uses of the drug, the possibility of reducing the suffering and debility caused by the disease greatly improved. But since the way malaria spread still was not known, no concerted attack could be mounted against it. Malaria began to recede spontaneously from areas where winters were hard, houses tight, cattle penned, and frontier conditions of shortest duration—and none of these were southern characteristics.[4]

Even during the first half of the nineteenth century, the sickliness of the region relative to the rest of the country was often remarked. Significant was the disease record of the army. Members of one of the few organizations of the time that could be termed national, the Regulars enjoyed (at least after 1832, when rigorous examinations were instituted) the care of more skillful doctors than the average citizen could employ. Troopers experienced life in northern, southern, and western posts. Though malaria and scurvy often made their lives on the western frontier grim, conditions in the southern posts, even those long settled, were worse still. Outstanding was Baton Rouge, the unhealthiest post in the nation, which recorded, year after year in the 1820s, over one-fifth of the total deaths in the peacetime army, most from dysentery, malaria, and yellow fever.[5]

Complicating the effects of climate was the rural population's truly medieval unconcern with wastes of all sorts. In a recent and excellent study of antebellum Virginia, medical historian Todd L. Savitt has portrayed the life of plantation slaves in a world of agrarian insouciance

Savitt, Medicine and Slavery

where "even privies were infrequently used" and refuse of every description was hauled to the fields as nightsoil. Though admirable as evidence of the growth of intensive farming in the Chesapeake, this practice had unhappy side effects in the infection of the soil with intestinal parasites and the growing plants with various pathogens.[6]

Southern cities had their own problems. In the early nineteenth century a city in the South implied a navigable river, or the sea, or both. Even by the relaxed standards of the time, port towns were ill-cleaned, and low-lying urban areas like Foggy Bottom in Washington, D.C., were notoriously malarial. The ports were most notable, however, as the foci of imported, epidemic disease, which was seeded by ships, sailors, and immigrants, and sustained once it began by large, compact populations. The contrast between town and country disease patterns was brilliantly stated, indeed overstated, by Dr. Josiah C. Nott of Mobile:

Yellow fever is generated in crowded populations, perhaps exclusively; while bilious fever on the contrary, is the indigenious [sic] product of southern soils. In fact, there would seem to be something antagonistic in the causes of these diseases. Generally, along the southern seaboard, when the forest is first levelled, and a town commenced, intermittents and remittents spring up, and in some places of a malignant, fatal type. As the population increases the town spreads, and draining and paving are introduced, yellow fever, the mighty monarch of the south, who scorns the rude field and forest, plants his sceptre in the centre, and drives all other fevers to the outskirts.[7]

The major epidemics of the antebellum years were cholera—a nonsectional disease that attacked North or South wherever its water-borne and filth-borne organism, *Vibrio cholerae,* might be carried—and a once national menace that became sectional, yellow fever. All major cities— Norfolk, Charleston, Savannah, Mobile, even inland Atlanta—suffered epidemics of yellow jack. New Orleans, the great port of the Southwest and the largest city south of Baltimore, became a frequent victim and, in part because of yellow fever, in part because of cholera, and in part by reason of its rich assortment of endemic ills, the unhealthiest city in the United States. It suffered major outbreaks of yellow fever in 1796, 1817, 1832, 1835, 1837, 1839, 1841, 1847, 1853, 1854, 1855, 1858, 1867, and 1878. The epidemic of 1853 was the worst, killing some 9,000 people. As this list suggests, the most dangerous period for the southwestern cities and towns came after 1830, when the population of the river ports burgeoned as nonimmunes flocked in from the North and Europe. It was the tragedy of such places that population growth, by provoking widespread and fundamentally incurable epidemics, contained its own automatic check and nemesis.[8]

In the toll of urban disease, ignorance, laissez-faire, and the general weakness of governments all played a role. City governments of the period, though gaining new functions, were closely identified with the merchant class, and existed primarily to protect property and preserve the established social order. Businessmen did not, of course, favor disease; they did, with good reason, distrust contemporary means of preventing it. The only true specific was flight. In New Orleans, a vast transient population concluded its business in the spring and left town at the onset of the sickly season—an annual migration which rumors of an epidemic turned into panicked flight. Newspapers and the city fathers meantime typically maintained a discreet silence for fear of injuring business. ("It spoils the sale of . . . goods," said Honoré Grandissime in the George W. Cable novel.) As the disease took hold, journals at last began to print the news that was on everybody's lips. Neighboring jurisdictions and towns on main escape routes attempted to exclude refugees, and even refused to handle mail and goods which had originated in the stricken city.

Reinforced by the onset of cholera, a science of public health had begun to emerge, especially in Europe, but aside from efforts at quarantine and brief cleanups following usually upon epidemics, few effects were seen in the South. There were indeed differences between towns, Savannah opening in 1817 a remarkably ambitious campaign against the wet cultivation of rice in its vicinity. The residents of New Orleans, on the other hand, became known for their indifference to elementary precautions and their bland insistence that their town was a very healthy one. Here civic pride, indolence, the wants of commerce, and the belief that only newcomers were in danger made a deadly combination. Since yellow fever could be experienced in mild as well as acute form, and since the disease was so frequent a visitor, it is easy to understand why longtime residents were more likely than immigrants from Europe or the North to have acquired immunity. Additionally, the blacks enjoyed an inherited resistance that made them no less likely to be taken ill but far less likely to die. But variant strains in the yellow fever virus sometimes brought surprises. In 1853 large numbers of residents died, as they were to do again in 1878, along with substantial numbers of blacks. The view seemingly held by some citizens that epidemics were "providential instruments for keeping down the number of undesirables in the population" was of limited validity.[9]

Yet in many ways the response of southern towns to epidemic disease was entirely respectable in the context of the times, and that of some individuals was heroic. Local governments labored to control disease among the poor; free inoculations against smallpox were sometimes provided the indigent in an exercise of enlightened self-interest. Quarantines were intermittently attempted despite the complaints of shippers and the doubts of many doctors who believed that illness was the result of climatic conditions; despite, too, the frequent failure of quarantines to prevent epidemics. The compassionate spirit, embodied in groups like the Howard Associations and New Orleans's state-assisted Charity Hospital, was common in afflicted cities. The fundamental problem in fighting disease was the bedrock fact that so little was known about its causes—and for that, no one, not even the despised medical profession, could reasonably be blamed. The effects of disease were an environmental calamity far more than a consequence of social or moral failings.

To assess the impact of disease upon the growth of southern cities, or the section as a whole, would be a daunting task, even if possible to carry out. Did the regional reputation, for example, play a part in the fact, so curious to twentieth-century eyes, that the flow of vacationers was from South to North, almost never the other way, or that northerners were by and large profoundly ignorant of the South? It is certain that epidemics imposed a heavy tax on the commercial life of the port cities, ending while they raged all forms of trade—ocean, coasting, and river; export and import; domestic and foreign. Even the rumor of an outbreak could temporarily disrupt trade, and internal unemployment and destitution followed. The towns were left at a disadvantage against northern competitors, once the railroads had made much of their market region no longer captive to their misfortunes. Even on a captive market, epidemics imposed a burden, as businessmen charged higher prices afterward in an attempt to recoup their losses. Insurance companies exacted higher premiums or declined to write policies altogether. Josiah Nott noted that "from a half to one per cent more is demanded on southern than on northern risks [insurees]," which he attributed to northern "exaggeration" of southern disease.[10]

Moreover, immigration of the most literate and propertied outlanders and their patterns of settlement after arrival were affected. The century saw steady increases in the volume of knowledge, as literacy spread and newspapers and books multiplied. In consequence, it became less easy for city fathers to conceal or local newspapers to lie successfully about epidemics. Immigrant guidebooks, especially those written by Germans with experience of America, often contained candid accounts of disease in the Mississippi lowlands, influencing their readers to turn toward higher, drier, and cooler areas in Texas and upland Arkansas. In the lowlands, wrote Traugott Bromme in his popular and much-reprinted guidebooks to the Americas, "the land expires pestilential mists," creating a climate that "is for new settlers very unhealthy." Louisiana was *"äusserst ungesund"* (extremely unhealthy) and New Orleans *"höchst ungesund"* (supremely unhealthy).[11] Locals were alive to the problem. Warning that "no one abroad gives credit to the oft-repeated assurances of the salubrity of the city," the president of the Medical Society of the State of Louisiana declared in 1851 that "the city has been characterized abroad as a great Golgotha, and signalised for its perennial pestilence." That flagrant booster of Louisiana and the South, *DeBow's Review*, often deplored what it viewed as the errors of northerners and immigrants in thinking the South as a whole, and expecially the lower Mississippi valley, to be unhealthy spots. To its credit, *DeBow's* also published the work of New Orleans's own medical statisticians, which tended to show that, even in good years, the city's death rate could be expressed in multiples of those of port cities in the North and Europe.[12]

Too much may, therefore, have been made of the city's absolute growth in the antebellum years. Outsiders did come, risking death for wealth; prosperous immigrants used the city as an entrepôt because there was no other in the Southwest, and pressed hastily on to safer regions. Poor immigrants like the Irish endured the harrowing of disease and afterward built up their numbers from the surviving remnant of natural immunes. The counterfactual question of what New Orleans or the lower Mississippi valley might have become had they not been the nation's unhealthiest region is pertinent, however, if perhaps unanswerable.

To be sure, New Orleans and southern ports in general had many other disadvantages. In their basic economic role as importers and exporters they suffered by their geography, which made them relatively remote from Europe. They were troubled by obstructing bars of sand and mud that formed where alluvial rivers met the ocean. In the alluvial valley they faced competition for midwestern goods from a growing northern canal and railroad system. Their long-range failure to grow competitively may have been inevitable, regardless of the geographical advantages that had created them in the first place. But epidemic disease and early death told against cities already at a disadvantage, and helped to define their parochial future.

Only less poignant than the waste of human lives was that of forests and the animal life they contained. Here events took a largely random course, governed by laissez-faire and limited by lack of population and industry.

As in the past, such population as there was had an impact on the southern forests out of proportion to its numbers because of the extensive and shifting character of southern agriculture. There were other uses for trees. In the country, fencing continued to absorb excessive wood to exclude wandering cattle. (There is evidence that in the Virginia Tidewater, some planters, by changing the type of fences they built, were deliberately conserving their shrinking forests.) Wood-built cities provided a market for the regional lumber industry. Railroads were often granted the timber that grew along rights-of-way. On rivers, much wood was consumed by the boilers of steamboats. The national lumbering industry, on the other hand, though its production reached one billion board feet in 1840, remained centered in the North until the Gilded Age. The great commercial assault on southern timber was still to come.[13]

North Carolina continued to be the national center of the naval-stores industry, outproducing all other states combined. The longleaf pine forest, spread by Indian burning and exploited for tar and turpentine by black laborers at white behest, might already have been considered a human domesticate, if not another in the list of southern staple crops. If it was a crop, however, the forest was one that grew slowly, maturing over decades rather than seasons. The organization, time, and knowledge required to harvest such a crop while replacing it was entirely beyond the antebellum South. What the age discovered instead was more uses for the trees, primarily for turpentine as a solvent and illuminant. Increased production by crude extraction methods meant the death of many pines. Additionally, a pine-borer infestation during 1848–1849 destroyed hundreds of thousands of trees. Hence the 1850s saw increased production in the lower South, particularly around Mobile.[14]

There were some contemporary echoes of the Broad Arrow. At the end of the eighteenth century efforts by the federal government had begun to preserve timberlands for the navy. In 1799 the government purchased Grover's and Blackbeard's islands off the coast of Georgia for $22,500. In 1807 Congress legislated against trespassers on public timberlands, and in 1817 authorized the president and the secretary of the navy to reserve public lands containing live oak and red cedar trees. Three islands in Lake Chitimachee, Louisiana, were withdrawn from sale in 1820 for the sake of their oaks. This policy continued until the 1840s, by which time some 250,000 acres of forest in Alabama, Georgia, Florida, Louisiana, and Mississippi had been set aside. A forest experiment station was set up in 1820 on Santa Rosa Peninsula, Florida. But it was easier to legislate than to prevent trespass, and illegal logging as well as squatting was common, despite a new law against timber trespass passed in 1831 and strengthened in 1859.

By this time Congress had already reversed itself and gradually began to dispose of the government's reservations. The advent of iron warships put an end to the status of shipbuilding timbers as strategic materials. Yet the antebellum years marked an important first for the national government in attempting to reserve and protect forests. As a political and legal precedent, the effort transcended its pragmatic and strictly limited aims.[15]

Neither industrial exploitation nor governmental efforts to protect set the dominant pattern of forest use, however. Far more important than either was the destruction of forests for cropland. This agricultural assault meant that areas unsuited to the hoe were likely to be spared, unless they happened to lie along streams that made their timber readily marketable. Thus the Appalachian hardwood region was so rugged and inaccessible that early lumbering was minimal and the farmer seldom intruded. The primeval forest was formidable—some 325 billion board feet by one estimate—and trees grew to extraordinary size. Settlers confined their clearing mainly to arable spots along the larger streams, though some lumbering occurred from the time of the earliest settlements. Supposedly the first sawmill west of the Alleghenies was set up in 1776 in what is now West Virginia. The Parkersburg Mill Company was launched in 1825. From 1830 on, some water-driven mills were at work along the larger streams, forty or so in the West Virginia area alone. But many hill-country streams were too small, crooked, or swift to move logs efficiently. Only after much of the northeastern forests had been cut did large-scale commercial logging come to the region, and only with the advent of the railroads were areas far from the rivers made accessible to exploitation.[16]

Similarly, the pine barrens during antebellum times were not much exploited by lumbermen or farmers. Except in the river bottoms the soil was of little value. Stockmen used the region, grazing cattle in grassy glades kept open by burning. In the Old South (as later in the New) the pine country was a summertime refuge for dwellers from the coasts and cities seeking escape from heat and disease. The regional legend of the healthiness of the "pine air" reflected the real absence of infected water supplies and human carriers of exotic ailments. Though agriculturists scorned the soil and lumbermen tended to despise the trees, visitors, especially foreign ones, sometimes found an unexpected and compelling beauty in the pinewoods: "The eye was bewildered," wrote a British traveller in 1838, "in a mass of columns receding far back and diminishing in the perspective to mere threads, till they were lost in the gloom. The ground was everywhere perfectly flat, and the trees rose from it in a direction so exactly perpendicular, and so entirely without lower branches, that an air of architectural symmetry was imparted to the forest, by no means unlike that of some gothic cathedrals."[17]

Between the barrens and the mountains stretched a mixed region, typified by the Georgia Piedmont. Though pines were more common west of the Atlantic-Gulf divide than east of it, the general pattern aboriginally was of pines giving way gradually to mixed and then to hardwood forests as the land rose. In 1791 Bartram described the climax forest of the region as dominated by oaks, hickories, and pines. By their usual practice of cutting over much land so that fields could be shifted as they were depleted, settlers transformed much of this land into a scrub dominated by the quick-growing pines.[18] West of the Mississippi such land was covered with a similar mix of oak, elm, hickory, maple, hackberry, walnut, and locust. Because settlement reached it later, large-scale exploitation of the Arkansas country had only gotten under way in the decade before the Civil War.

The swampy lowlands of the South were immensely varied. The eastern wetlands formed a double belt along the coastline, salt marsh in contact with the sea, and brackish or fresh-water swamps fed by tidal rivers behind the coastal ridge. Rapidly growing vegetation fed plankton and shrimp, which provided the base of the food chain. The abundance of food drew an abundance of species. Anadromous fish entered the streams of the marshes seeking the ponds of their birth. The movements of the tide supplied an energy subsidy, repeatedly washing through the sponge-like channels and sweeping detritus in and out. This created a literal counterpart to Walt Whitman's image of the sea as a cradle endlessly rocking, for the movement of nutrients made a nursery where the young of many species could feed and grow. Along the Atlantic flyway migrating birds exploited the marshlands as wintering areas or for stopovers on long flights between the far North and the tropical South. Human fishermen and hunters followed the same magnet.

Around the tip of Florida, where mangroves stoppered the overland flow of water through the Everglades, and along the Gulf coast, the scenes were similar. Wetlands formed around the exits of alluvial rivers, the inefficiency of the lower channels causing a rhythmic overflow in times of high water that shaped marshes such as those of the Pearl River and the lower Atchafalaya. Here, and in the immense salt- and fresh-water marshes of the Louisiana coast, the tidal energies, warmth, and seasonably varying salinity gradients helped to sustain a profusion of living things.

Wetlands were not ony to be found along the coast. Inland swamps existed along all alluvial rivers, functioning as reservoirs that absorbed and stored water during the spring floods and released it as the rivers fell. (How essential this was, human occupants of the river basins would learn when they cut off the lowlands with levee lines.) In the sloughs and back-waters existed many species both permanent and transient. The birds were perhaps the most spectacular of the many forms of life. In the early nineteenth century, a traveller saw among the buttonwood trees of a Mississippi bottom "whole flocks of screaming parrots [parakeets] greedily feeding," and the change of seasons brought flights of ducks, geese, swans, and cranes down the Mississippi flyway to wintering grounds in Louisiana and beyond.[19]

For such regions the Old South found little use other than hunting for pot or market, fishing, and some logging. In the lower Mississippi valley cypress was the wood most sought from the swamplands; abundant, twice as durable as pine, and well fitted to resist damp, cypress was important in shipbuilding, and in New Orleans was widely used for houses as beams, as flooring, and particularly as roof shingles. Cisterns, fences, and coopers' staves were also made from the soft, sweet-smelling wood. As a consequence of the federal Swamp Lands Acts of 1849–1850, expanses of the best cypress land in the valleys of the Mississippi, Red, and Atchafalaya rivers passed via frauds by state agents and surveyors to men of political influence. Much was sold to lumbermen. Efforts by the state to prevent illegal cutting apparently had little effect, as did federal efforts, though gunboats were sent to the Atchafalaya and the Calcasieu rivers, and some logs cut from public lands were confiscated. Despite the energy and ingenuity of surveyors and loggers, however, the inaccessibility of many swamp areas delayed the major assault on the cypress stands until after the Civil War.[20]

East Texas woodlands were exploited not only by farmers and planters in the immediate neighborhood, but by lumbermen aiming to supply the settlers who were pushing tentatively into the treeless prairie to the west. East Texas forests were about half pineland, with the usual transition to hardwoods in the north—mostly oaks, with the customary mingling of hickory, elm, sweetgum, etc. Swamp and bayou lands held the much-valued cypress as well as other wetland varieties. In 1819 a sawmill was built in Nacogdoches, and some exporting was done to Mexico. Rapid settlement of the northeastern region from Nacogdoches to the Red River accompanied a variety of developments that meant new pressures on the forest: the growth of a local iron industry, construction of a railroad in 1858, and, to the south, the founding of scattered settlements and the introduction of many cattle. Early cutting was wasteful in the extreme. Trees were sawed off at waist height and the stumps left to rot; cypresses logged at high water, when the wood was easier to float, might be cut twelve to twenty feet above root level. In neighboring Arkansas lumbering had become the leading industry of the state by the time of the Civil War. In neither state, however, had more than a beginning been made in an industry whose maturity was deferred until later in the century.[21]

The years between the Revolution and the Civil War had brought extensive changes to southern forests. Outside the naval-stores region of North Carolina, the heaviest such impact probably occurred in the Piedmont regions. Here the demands of agriculture resulted in the replacement of great areas of a predominantly hardwood forest with old-field pines, and in the process injured soil, water supply, wildlife, and even topography, for much of the erosion which took place must have resulted from runoff, not over plowed fields, but over cutover or fallowed land.

Yet, while sawmills and the naval-stores industry spread and some serious damage occurred, by the time of the Civil War the whole South had few more sawmills, and those with a lower capital output, than the state of New York alone. The agricultural fixation of the South, spreading ruin in one place, protected or at any rate ignored other woodlands; distance from rivers, a low population, rugged or nonarable land could still provide adequate protection in a section which held its frontier character until the Gilded Age. The spirit of the Old South was seemingly quite willing to cut down its forests, almost to the last tree; but performance was incurably weak.

For the wildlife that filled the southern woods, wetlands, mountains, and prairies, the period before the Civil War was a hard one, though not yet the hardest. No innovations appeared in efforts to control the exploitation of game, and actual legislation may have been rarer than in the colonial past. Historically protection has been associated with aristocratic pleasure, sovereign right, bureaucratic regulation, or scientific thought. To none of these was the new democracy a friend.

Some regulation was attempted. Legislation patterned on colonial models forbade burning of woods and placed the usual bounties on wolves, wildcats, and other unfavored species. Alabama and Florida sought to establish a burning season in the spring, while Louisiana in 1845 sought to prevent the burning of "prairies" (probably the *prairies tremblants*, or marshes). Efforts were made by Georgia to regulate the cutting of trees, and various states legislated against blocking streams by throwing cut trees into them. The dangers of hunting with torches at night led Alabama to impose double damages for the killing of domestic animals under such conditions. Poisoning fish was forbidden in Arkansas; Florida too sought to regulate its fisheries. Arkansas, Kentucky, Florida, and Tennessee legislated against wolves.

Deer were a staple of the rural southern table, especially in the first period of settlement, and remained an important source of food, especially in the fall and winter, in areas where settlement was not too heavy. As the Indians were expelled and cattle replaced deer in the leather trade, the species probably recovered their numbers. In a region where 87 percent of the land was uncultivated, there was ample cover, and the relatively mild winters also favored their increase. The spread of white settlement, however, meant new danger, and the deer's loss of economic importance was the cause of much diminished concern for their survival, and so of laws aimed at protecting them. Pursued with relish, and hunted by men armed with ever improving weapons and no scruples about methods, deer may have been killed at an annual rate of one and a quarter to three million, from a total population of six to ten million. Such pressure, when combined with ever diminishing habitat, by the end of the century would reduce the species to about a million animals in the region.[22]

Hunting for sale was carried on wherever possible. Despite the lack of quick transport, game animals hung in the markets of Charleston, New Orleans, and many smaller cities. Overall, the spread of agriculture, the doctrine of free taking, and the absence of conservation made for serious losses, especially in the older regions. In his well-known *Carolina Sports*, William Elliott of South Carolina recounted not only the joys of hunting but the increasing depletion of outdoor life in his neighborhood near Beaufort. "No living man," he wrote, "has seen a wild buffalo in our confines. It is a rare thing to encounter a panther. . . . The wolf is almost extinct; the bears are fast diminishing in numbers; the deer, though still numerous in given sections, are visibly thinned; and it is only the smaller animals, such as the foxes and wild-cats, which are still numerous, or whose diminished numbers have not been made the subject of remark."[23]

The prime causes, according to Elliott, were market hunting and the destruction of the forests—that of the river swamps by rice culture and of the highland forests by cotton. Efforts to preserve game, even in his own long-settled neighborhood, where property rights might be presumed to be strong, appeared to be hopeless. Elliott made the attempt, denying his overseer the right to hunt, with the result that his neighbors "took a malicious pleasure in destroying the game which a proprietor had presumed to keep for himself." Acquaintances who owned islands and had set bailiffs to exclude poachers found the worst poachers in the bailiffs themselves. "The right to hunt wild animals," he concluded, "is held by the great body of the people, whether landholders or otherwise, as one of their franchises; which they will indulge in at discretion; and to all possible limitations on which, they submit with the worst possible grace!"[24]

Elsewhere the range of the bison was again restricted, as the animals were exterminated east of the Mississippi and nearly, if not quite, destroyed in the alluvial valley, as well. Quail and turkey remained plentiful, but passenger pigeons were seriously depleted by comparison with the preceding century, and parakeets declined, not only killed by farmers indignant at their fruit eating, but captured to be kept as cage pets or shot for their feathers. Yet no species appears to have become extinct. The age of extinction, like that of conservation, lay ahead.

The limitations inherent in the Old South's attempts to exploit its resources showed clearly in its efforts to overcome natural obstacles to commerce. Viewed as avenues of transport, southern rivers had distinct advantages and notable failings. Except in northern Virginia, the highlands, and the border regions, they seldom or never froze. The fanlike pattern of streams running off the central highlands provided useful routes for inland products to the sea and beyond. The Mississippi delivered the wealth, as well as the floods, of a continental basin to New Orleans. But the alluvial rivers were exceedingly variable streams; freshets on the Chesapeake rivers could bring sudden rises of forty feet or more, and throughout the coastal plains unstable, twisting channels shaped by the meander process made navigation a perilous art. Building railroads to supplement water commerce was one answer to the problem; another was to "improve" the rivers themselves by removing obstacles, straightening channels, and building levees.

Recent quantitative studies have indicated that the achievements of the Old South in building railroads were more than respectable, at least by comparison with contemporary European nations. Land grants by southern states were often generous, demonstrating that southerners, despite all differences, were Americans at heart. But the lines were mostly short and were intended to bring staples from upland regions to the ports. In 1848, New York, Pennsylvania, and Massachusetts had more miles of railroads than all the slave states combined. Major railroad impacts were essentially postwar.[25]

More typical of the times were efforts to transform the rivers and to supplement natural by manmade waterways. The financial support given such projects by the southern states varied widely, but in general was vigorous up to about 1836, when it became apparent that heavy costs, local politics, and bad engineering had caused many serious losses. One canal, at Louisville circumventing the falls of the Ohio, was a notable success; many others, despite heavy infusions of state money and some federal aid, were not. Federal intervention was an important factor in meeting needs for both money and skill, but was limited by constitutional strictures and by the quarrels of competing states and sections. Under the program of national development pursued by the Young Republicans after the War of 1812, the federal government provided subsidies from land sales to Indiana, Mississippi, Alabama, and Missouri to build roads and canals. In the 1820s special grants were earmarked for favored roads and canals, and technical assistance was made available to the Baltimore and Ohio Railroad. The help given by army engineers was a subsidy of knowledge no less important to new projects than those of land and money, for until 1824 West Point was the nation's only school of engineering.[26]

In later years the South's growing localism helped to impede federal action throughout the country. John C. Calhoun's conversion from proponent to enemy of federal support of internal improvements was striking; but two other southerners were of more importance to the evolution of policy. Chief Justice John Marshall, in his *Gibbons v. Ogden* decision of 1824, declared the primacy of federal power in assuring free use of navigable waterways; his decision opened the way for Congressional action on the Mississippi and, it seemed, on many other rivers. But in several of his vetoes—of the Maysville Road, and of rivers and harbors bills—President Andrew Jackson forced a new and more cautious approach upon the nation. Though full of paradox on the issue, his administration marked a turning point after which such projects increasingly fell victim to sectional and state jealousies veiled in constitutional scruples. Institutionally, the country lacked any national organization willing and able to impose consistent policy, and it was perhaps more than anything the inconsistency of federal action that spelled the end of an effective national civil works program. Erratic and undependable improvement was worse than none at all.[27]

On these issues the South was far from united, the Southeast generally against federal action, or at any rate badly divided, the Southwest in favor of it. Perhaps the localistic, cautious and nay-saying aspect of the South was merely more representative of the nation than the other. In any case, the alluvial states of the Southwest found themselves facing a continental river system with local and divided forces. Again the will to levee, develop, and improve was strong but the means insufficient. Local treasuries were often at low water when the Mississippi and its tributaries were the reverse. Levees to be effective had to be skillfully designed and built, had to be continuous, and had to be maintained by a power with access to large and dependable supplies of money. What local authorities had was ingenuity and hard work, which proved insufficient. An important innovation, apparently pioneered in Mississippi's Yazoo basin in the 1840s, was the levee district, a legal construct with powers to tax and issue bonds to finance levee construction and maintenance. The embankments which resulted stood up to severe floods in 1861 and 1862, only to suffer later from wartime neglect and the operations of General Ulysses S. Grant.[28] More commonly levees were low, weak, and often sited wrongly. Nobody as yet had any very useful idea of just how much an effective system would cost, in part because nobody yet recognized how high a levee system, by contracting the river, would cause its floodline to rise.[29]

Baffling too was navigation. The collapse of forested banks added a yearly quota of snags to alluvial streams. In 1819 a flatboatman on the Mississippi witnessed "horrid sinkings of the bank, by each of which not less than an acre of land had fallen in . . . with all the trees and cane upon them." In the reach between the Arkansas and St. Francis rivers he saw a boat "which hung upon the trunk of an implanted tree, by which it had been perforated and instantly sunk." This was cause and effect. Taking heart from Marshall's decision and the Commerce Clause, the national government tried to do something about navigation. From 1824 until 1854 both soldiers and civilians such as Henry M. Shreve labored to remove snags and to break through the Red River raft and similar obstacles. Shreve also experimented with a new artificial cutoff of one of the Mississippi's meander loops, with rather unhappy results. Officers surveyed and mapped the rivers of the West, and uniformed engineers worked to keep open port cities, as Robert E. Lee did at St. Louis and Pierre G.T. Beauregard did at New Orleans. But the results were, at best, to alleviate the continuing dangers to a navigation which often seemed to move in spite of the rivers. Efforts by the western states to obtain greater federal help ran into the old sectional obstacles, though not always those of North and South. In November 1845 the South-Western Convention at Memphis advocated federal responsibility for promoting navigation along the "great thoroughfare" of the river. Federal action against floods was also recommended, by making land grants to aid the states.[30] This was an ingenious suggestion.

Forests, land, and rivers, reclamation and flood control were all involved in the Swamp Lands Acts of 1849–1850. Louisiana, Arkansas, Mississippi, Alabama, and Florida were the five southern public land states. Federal grants of such lands would make it possible to aid land reclamation while enriching local magnates—an irresistible combination of worthy goals. Severe floods in 1849–1850 helped put pressure on Congress to act. In March 1849 the lawmakers granted Louisiana all the publicly owned swamp and overflowed lands in the state which were unfit for cultivation, provided that the money obtained from their sale be used to levee and drain them. Since the role of the swamps as reservoirs in times of flood was apparently not widely understood, lawmakers may not have been aware that leveeing the swamps would, among other things, increase flooding. But even the few who understood this need not have worried; there was never much danger that leveeing and reclamation would come out of the acts.

In the same year a Florida senator introduced a bill to turn over the Everglades to the state to be drained. Congress rejected this proposal, but in 1850 donated all swamplands in the states of Alabama, Arkansas, California, Florida, Illinois, Indiana, Iowa, Michigan, Mississippi, Missouri, Ohio, and Wisconsin to the states, with the same proviso. Under these acts the lucky states could follow the surveys of the General Land Office to determine the character of the land, or supply their own evidence that it was wetland. Most chose the second option.

In consequence, much of the sixty-three million acres ultimately disposed of was timberland, including valuable stands of Louisiana cypress; much was not wetland at all. Appointing their own surveyors and commissioners, the states presided over a process by which useful land passed into the hands of influential individuals and lumber, railroad, canal, and drainage companies. A substantial part of Florida's twenty million acres lay in the northern part of the state rather than in the southern marshes. Little reclamation was carried out anywhere. But local fraud, like local action generally, had severely limited effects upon the environment. Honesty, organization, and efficiency were far more likely to produce great changes, and those qualities belonged more typically to the institutionalized future.[31]

The science that could guide either exploitation or conservation, and which in practice has been indispensable to both, was pursued sporadically in the Old South. Technical innovation had been decisive in giving the row-crop empire its historic form; the low estate of medicine was equally important in the continuing toll of disease that harassed and retarded the region. Southern scientists favored biology and earth science, included in their number men of national reputation, and were limited in numbers far more by ruralism than by slavery. Perhaps the most typical work of applied science was the survey, certainly the most logical in view of the state of knowledge. On the other hand, scientific speculation, medical and otherwise, tended toward rationalizing the slave society—racism with calipers and a microscope. Simpler collectors of plants and animals added significantly to scientific knowledge, and artist/scientists in the Catesby mold continued to appear. The South found in Audubon, an indestructible waif who adopted the nation as a parent and falsely claimed southern birth, one of its supremely interesting birds of passage.

In attempting to survey and control their geological resources, southern states had a mixed record. Surveys were initiated by the Carolinas in the 1820s. During the late 1840s and 1850s Kentucky and Alabama made the professor of geology at the state university responsible for such investigations "as the Legislature should meagerly provide for."[32] Explorations of this nature were meant to promote private exploitation, not reservation for controlled use. During that classic decade of laissez-faire, the 1840s, most states which had practiced reservation abandoned it as "uneconomic." In this, Texas swam against the national tide. In 1840 the Republic expressly retained Mexican laws reserving "mines and minerals of every description" under the Spanish doctrine of a special sovereign right in such deposits. The reservation lasted only until 1866, but for a time, at least, made an exception to the commoner practice of unrestricted private right.[33]

Serious work was done by the engineers Andrew A. Humphreys, Henry L. Abbot, and Charles Ellett in their surveys of the Mississippi River. The two major investigations, though by no means free of technical flaws, were extended and thoughtful efforts to comprehend the working of the great alluvial river. In their *Physics and Hydraulics of the Mississippi River*, Humphreys and Abbot introduced a wealth of statistics, sometimes dubious, while Ellet's approach in *The Mississippi and Ohio Rivers* emphasized prophetic insights, sometimes mistaken. Both covered a remarkable range of possible means for transforming the Mississippi into something closer to the human heart's desire. Levees, reservoirs, outlets, and cutoffs—the proposals and scattered attempts of a century and a half of occupation by European man—were reassessed. Humphreys's and Abbot's suggestion that levees alone could restrain the river portentously entered engineering consciousness. Under the antebellum system neither their survey nor Ellet's could have led to action (Humphreys's and Abbot's work, in any case, was not published until after the war had begun) but both remained as the prewar era's guide and *vade mecum* to Mississippi valley politicians, engineers, and planners of the Gilded Age.[34]

The higher reaches of southern science participated in the intellectual movements of the times while often bending them toward the needs of an increasingly embattled section. Many of the South's physicians were at work providing medical justifications for slavery on the basis of a melange of observed and imagined differences between blacks and whites, with the aim of showing that the slave might reasonably be kept in bondage because he was as close as possible to being a different species from his owner. As "southern medicine" sought to provide a rationale for a social order grown odd, archaic, and defensive, it became a harbinger of trends toward racist exclusivism that grew more powerful and widespread later in the century, when the western world at large found itself in a position somewhat analogous to the South's—entrenched rulers over a planetary empire of the dark skinned.[35]

Considerably more innocent and of more lasting value was the work of the naturalists. At its upper level, this group included scientists of real distinction. The first half of the century was a time of change that saw the rise of the academic specialist and the transformation of a hobby into a profession. Many of the South's naturalists, however, were collectors, men of the frontier, or persistent travellers. They had, of course, contacts with the centers of culture—not only Europe and the older coastal cities of the North and South, but the developing inland centers like Lexington, Kentucky, at whose Transylvania University Daniel Drake set up the medical faculty, and where natural history was taught for a time by the gifted eccentric, Constantine Rafinesque.

In a field which by now has received so much scholarly attention, even a list of a few major names may be needless. Yet it would be hard to speak of the southern wilds without noting the influence of Jefferson, whose enthusiasms were an enduring source of southerners' interest in their own wilderness. Stephen Elliott's *Botany of South Carolina and Georgia* was accounted distinguished work, and the Charleston clergyman John Bachman contributed much to Audubon's *The Viviparous Quadrupeds of North America*—and still more to his family, for his daughters married Audubon's sons. Henry W. Ravenel, John E. Holbrook, and Matthew F. Maury won fame in natural history and oceanography. As in the eighteenth century, travellers provided much of the labor and insight that went into an expanding knowledge of the southern wilds; among the most distinguished were André Michaux of France, Scots-born Alexander Wilson, and Thomas Nuttall, a Briton who spent much of his life in America, living at Philadelphia and later teaching at Harvard, but travelling widely. Among the native collectors, John Abbot, George Engelman, and Ferdinand Lindheimer contributed information and specimens to the "closet naturalists" at Harvard and the Smithsonian.

As for Audubon, his contribution was unique, transcending any one section, transcending too the tradition he worked in. An artist who learned science, rather than the other way around, he painted in a surprising range of styles and manners. Some of his creatures were stiffly heraldic and seemed to demand a shield to pose upon. Some paraded against southern backgrounds painted by his assistants. Most—perhaps the best—displayed against a white background the elegant forms that Audubon seemed chiefly to delight in. On occasion the artist showed himself a Victorian anecdotalist, a Dickens of birds and beasts. Character filled the best of the portraits. The robin's domestic life supplied one of his most complex designs. He caught the fierceness of the mockingbird and showed it dramatically repelling a serpent. (He might equally well have shown it persecuting a tomcat). A marvelous portrait of the yellow-breasted chat equalled the great bird-and-flower compositions of China and Japan. It is true that melodramatic and contorted posturing (the latter forced in the case of the large birds by an unwise attempt to portray them life-size) sometimes marred his work. His later pictures showed declining energy, or perhaps interest. The animal pictures turned out by his workshop were rarely the equal of the greatest birds. Overall, however, he bequeathed a treasure from which anyone susceptible to art or nature may pick his own favorites and find others at the next perusal.

There was also a bloody and incomprehensible side to the activity of the naturalist pioneers. Hunters for food and sport as well as collectors, they killed with almost as little restraint as the woodsmen who surrounded them. Audubon remarked of the Carolina parakeet that he had "procured a basketful . . . at a few shots," and on a hunt for Canada geese exclaimed, "Oh, that we had more guns!" When one finds a collector forwarding to a museum dozens of specimens where a few would have done, the fact is borne home that even many of those who loved the wild, sought to know it, and lamented its transformation, treated its abundance also as a commodity, though in this case the wealth sought was of knowledge or at worst of a naturalist's modest fame. Again, like Audubon, many went hunting as a jogger takes his morning run, for mere exercise. But they left a legacy of useful knowledge—or useless knowledge, which may be self-justifying as an exuberance of the spirit. And Audubon left an undying image of the passing wild at a time when, though under pressure, it still survived almost in primeval form through much of the newly settled South.[36]

6

Exploitation Unlimited

FOR MANY SOUTHERNERS THE GILDED AGE WAS A HARDSCRABBLE TIME, AND their environment suffered accordingly. Poverty is no friend to natural resources, which typically are devoured piecemeal to sustain existence. Poverty plus a worsening disease environment severely impaired the region's human resources, as well, for the South's endemic ills probably became more widespread as well as more peculiar in the generation when medicine was transformed by Louis Pasteur and Robert Koch. Its epidemic yellow fever, though visitations became rarer, remained a regional trademark.

What happened to the South's natural wealth depended very largely upon the action of outsiders, who cut its forests, bought up its land, and financed its railroads and many of its nascent industries. Manufacturing, after a shaky start, grew rapidly in the 1880s, as southern textiles began to cut into a traditional New England industry. Cigarettes, fertilizers, and vegetable oils were also turned out. With few exceptions, however, industrial impacts on the environment remained of small account, as did southern production: in 1890 the South's entire output was worth less than half that of the state of New York.[1] Agriculture and the extractive industries were another matter. Forests and wildlife were brutally used, and the South produced cotton in extraordinary abundance, at a considerable cost to the land.

That uncommon flower of the age, the conservation movement, originated in the industrial North and the water-poor West. The South had some leadership to contribute, but played its most important role as an arena of conservation battles directed from elsewhere. Carpetbagging was the rule of the times, whether saving or exploiting was the goal.

Regional health and regional agriculture fundamentally shaped the environmental history of the time, each modifying the other. Epidemics were fewer than in the recent past, but severe when they did occur. Conservatism in public health continued to be marked, endemic diseases were not reduced in scope, and malaria and tuberculosis may have spread.[2]

State boards of health were set up throughout the South following the Civil War, but were ill-funded and possessed only limited powers to investigate nuisances and offer advice. Still facing outbreaks of yellow fever, southern states tended to view the proper activities of such boards as limited to quarantines. After the epidemic of 1878 had devastated New Orleans, Memphis, and other areas in the Mississippi valley, a short-lived national board was established with powers to investigate, advise, and enforce quarantine regulations. Effective work by the board in cleaning up Memphis and gathering vital statistics did not prove to be enough, however: board members lacked political influence and faced unrelenting opposition from local boards jealous of their turf. The organization went out of existence after only four years.

The medical profession itself remained confused, even though it was on the verge of great things. After a deade of resistance, American doctors began during the 1880s to accept widely the germ theory of disease. Sanitarians pioneered cleanups to prevent the spread of filth-borne diseases, such as cholera. More widely applied, their methods could have proven of great benefit to southern cities. Yet such methods were never more than marginally effective against mosquito-borne diseases.

Illness among blacks was a worsening problem. The upheavals of the time broke down the rudimentary health discipline of the plantations. Meanwhile rural blacks in large numbers and in utter poverty entered the towns. Even the limited sanitary arrangements of affluent neighborhoods stopped at the slums, where residents accustomed to casual country ways lived together in large populations well able to sustain many diseases. Blacks appear to have suffered greatly from tuberculosis. Enteric complaints of every kind were also common. Disease rates among the former slaves lent credence to an efflorescence of racist "science," including new medical studies which purported to show that excessive illness and crime among blacks resulted from physical inadequacies which in time would bring their extinction by natural causes.

But affluent whites did not evade disease either, and sometimes found ways to combat its symptoms which were as bad as illness. Postwar depression, aching war wounds, and the casual attitude of the age toward the use of opiates led to widespread addiction. "With the possible exception of the Chinese," a medical historian recently declared, "Southern whites suffered the highest addiction rate of any regional racial group in the country, and perhaps one of the highest in the world." Largely a response to cultural stresses, addiction also reflected in some measure the effort to control enteric diseases, for which opiates were commonly prescribed.[3]

In the port cities ships still arrived from abroad bringing cargoes of infection that were not on the bills of lading. Boats ascended the rivers, transshipping epidemics to the river towns. The abundant native vectors still bred in marsh, puddle, and jar, and the cities, by joining endemic to epidemic ills among dense populations, were unhealthier than the country. Yet conditions were not quite the same as before. Even in the Mississippi valley much of the heedlessness of earlier days had ended. At New Orleans quarantine regulations as ever proved easier to let slip than to enforce, and even when enforced, were carried out in ignorance of elementary facts regarding the cause and incubation period of yellow fever. Under these conditions the epidemic of 1878–1879 struck hard there and at Memphis. New Orleans now learned that it could no longer afford business as usual. Communities upriver made it plain that the city which drew upon the wealth of the valley had a compensating duty to protect it from epidemics. "New Orleans, at whatever cost," read a report of the Tennessee Board of Health, "owes it to all the territory bordering on the Mississippi to adopt the means to eradicate [yellow fever] . . . she is now looked upon with suspicion and distrust, her inland commerce embarrassed and liable at any time of alarm to be completely stopped." The reality of such warnings had already been demonstrated by many a "shotgun quarantine." Utter business paralysis during the outbreak led commercial leaders mainly in the domestic trade to organize the New Orleans Auxiliary Sanitary Association to promote public health, and the movement enjoyed some success even among foreign traders.

Upriver at Memphis, some historians have seen the same epidemic as a catalyst in far-reaching social changes. The fever, possibly a mutant strain, spread widely among groups such as blacks and natives who were supposedly immune. Prior to the epidemic, the town had possessed a heterogeneous population of the sort that had long been typical of the Mississippi valley. Widespread death among North Europeans (still the heaviest sufferers) may have resulted in large-scale decamping by Germans and a marked decline among the Irish as a vigorous political and cultural force. The city of the decade that followed was more heavily black, and its whites more often southern and rural in origin. Such changes may have marked the convergence of two peculiarities—the increasing uniqueness of the southern disease environment within the United States, and that of the South, with its small immigration, in American society. Whatever the impact of one epidemic upon one city, it is not unreasonable to suppose that these developments were in general mutually reinforcing.[4]

In the countryside the nature of southern agriculture continued to encourage malaria and parasitic diseases, while postwar changes probably increased the ills of malnutrition by destroying cattle and swine. Between 1860 and 1870 the number of horses in the South decreased by 29 percent, cows by 32 percent, and swine by 35 percent. Though corn remained for decades the region's leading crop, there is considerable evidence that the South became less self-sufficient in food production than before the war. To the burden of debt was added, in many parts of the lower South, an increasing need to pay for food imported from St. Louis, Louisville, Cincinnati, and other upper South and northern cities. Shrinking farm income during the 1890s added to the problem, imposing upon southerners a diet more monotonous, less balanced, and relatively more costly than in earlier times.

In regard to that continuing bass note in the symphony of disease, malaria, scattered evidence suggests that the number of victims increased. This is not unreasonable, since the dominant practices of southern agriculture remained much as they had been, except that the heavy loss of farm animals in the war presumably reduced still further the chance for "animal deviation," deflection of anophelines to domestic animals for their blood meal. It is true that malaria also increased in the North during the Civil War, as veterans of the fighting in the South returned to infect their own region. Additionally, endemic centers existed in some areas, including Staten Island, and epidemics occurred even in the early twentieth century as far north as Ithaca, New York. Such exceptions, however, did not alter the general picture of rapid recession for the disease in the North, while in the South a large, poor, widely infected rural population remained its major American reservoir. This peculiarity became more marked decade by decade.[5]

The Gilded Age ended in a medical calamity which was to have far-reaching and largely positive results. The onset of the Spanish-American War brought an influx of volunteers to southern training camps. Cobbled together by Congress under political pressure from the many who wished to serve, the volunteer army was ill-disciplined and ill-led. Many soldiers brought typhoid from the state camps where they had first been assembled. Then malaria broke out, as well. Through the summer of 1898 an epidemic gained ground whose makeup echoed earlier outbreaks in raw towns in the early settlements or on the frontier—a mix of malaria, typhoid, and dysentery. The army appointed a board of physicians under a Virginia-born officer, Captain Walter Reed, whose studies demonstrated the modes of transmission for typhoid and suggested proper means of control. The results of the study included an increase in knowledge, systematic application of prophylactic measures to the army, and a considerable boost to Reed's career. In the end, the third event would prove at least as important as the other two.[6]

These were extraordinary years for tropical medicine. During July

1898, in India, a British army surgeon completed his demonstration of the mode of transmission of malaria.[7] Reed and his colleagues would soon add discoveries on the transmission of yellow fever, as the Spanish war yielded the United States tropical possessions where the disease was a real, if often exaggerated, danger to its troops. At the end of the nineteenth century the South remained in a condition that was all too typical of the warm regions of the world. But a science of medicine had come into existence which was capable of identifying the specific causes of the great contagions. So enormous a revolution in the environmental relationships of the human species would soon react upon the South, as means of making the knowledge effective were slowly devised.

Much has been written on postbellum agriculture in the South. The system has been viewed mainly as an economic and social phenomenon, but the environmental aspects were fundamental. To summarize a much-told tale, the plantation survived the Civil War as a large landholding, sometimes in the hands of former owners but more commonly in those of new owners, private or corporate. Its onetime disciplined operation by a master or overseer commanding the labor of slaves was transformed. Some owners sought to preserve discipline by gang-labor, or at least by "through and through" plowing, working the whole plantation as a unit. But widespread tenancy tended to disintegrate the unity of the workforce; to the extent that they were able, freedmen preferred to regulate their own time and efforts. While plantation discipline waned, lack of capital compelled the creation of a jerrybuilt credit system under which farmers borrowed against the expectation of making a crop. Since cotton was the money crop in most of the lower South, merchants tended to loan on the staple, which some were also middlemen in selling. Banks were local and widely dispersed, enjoyed monopoly power in their own neighborhoods, and kept interest rates abnormally high. Obliged to grow cotton to get credit, and to work land with an inelastic force (his family), the farmer planted less food, which the merchant then obligingly sold him at a goodly markup.

In terms of both climate and custom, cotton production was rational. The tenant himself desired and needed to grow the fiber, for how else was he to face his creditors? Yet a great gap remained, and indeed grew wider, between the importance of cotton to the region and the nation, whose major export it was, and the rewards it brought to its growers. Between 1869 and 1899, the South produced some 198 million bales worth $8.5 billion. Meanwhile its tenants and sharecroppers were often described by census reporters as "without bread for their families" and "in worse condition than they were during slavery." The effect of the system was to reinforce a postwar boom in cotton while diminishing care of the soil and retarding diversification of crops. For this reason, the Gilded Age rather than the Old South was the true flowering of the Cotton Kingdom, the era of triumphant monoculture.⁸

Yet stereotypes have also been a southern staple, and the region has never fitted them very comfortably. As usual, immense variations existed among different regions of the South. Freeholds expanded even as tenancy was growing; in some areas agriculture diversified; in others, atypical before the war, farming showed northern patterns, with varied crops, including cover crops, and some dairying. Elsewhere ancient, indeed pre-Columbian, patterns endured. Viewing his own neighborhood, a Virginia historian of the late nineteenth century remarked that "a field of maize on the Powhatan, long before the first English explorers appeared . . . was almost the exact counterpart of the same field, planted with the same grain, three hundred years afterward by the modern Virginia farmer." Farming remained a most individualistic occupation, varying widely from field to field, with able and ingenious men standing out against bad practice wherever they lived.⁹

Agricultural improvement took varied forms. Use of commercial fertilizers became more common and more technically adept. The modern view that "land is only a vehicle" conveying calculated chemical inputs to the crop was energetically pushed by reformers. Perhaps these efforts served only to increase cotton production with unfortunate long-term effects for the farmer and his land. But diversification was also preached and practiced. In Virginia and North Carolina peanuts were becoming important, part of an innovative mixed agriculture that would in the long run supplant the old. During the 1880s, Georgia's peach trees, aided by P.J. Berkman's experiments, increased in number from 2.8 to 7.7 million, and orchards spread far and wide in the Gulf states generally and in Tennessee.¹⁰

The sugar country of south Louisiana was a surprising example of agricultural revival. The whole South found recovery from the war difficult; about twenty years were required in most areas to replace the physical capital that had been lost, and even the rapid growth of the 1870s did not bring complete recovery.¹¹ Yet an area hard hit by the war which had destroyed expensive machinery and dispersed the labor force began, in the face of recurrent floods which penetrated the damaged levee systems, a remarkable recovery. Between 1877 and 1910 drastic changes reshaped this branch of agri-industry, separating the manufacturing process from cane-growing, improving marketing techniques, and bringing the systematic use of fertilizers to the soil. Organized as the Louisiana Sugar Planters Association, producers in 1885 secured creation of the state Sugar Experiment Station, which began to apply scientific methods to improving production.¹²

In the same region rice culture benefited from an influx of midwestern settlers during the 1880s. The center of production shifted dramatically from coastal South Carolina and Georgia to Louisiana, and later to Texas and Arkansas, as well. Southwest Louisiana with its abundant streams and areas of retentive clay subsoil was eminently well suited to the staple. Spawned by railroad-building, commercial production of rice soon evolved, as that of sugar had earlier, into a sophisticated, large-unit agriculture. Problems of drought and salt-water infiltration (a persistent difficulty, given the landform) were met by use of pumps and canals, and after 1898 by pumping from wells. Efforts too were made to improve the traditional droving of half-wild range cattle by introducing purebred stock and cultivated grasses into the region. Towns sprang up, drawing income from agriculture and logging; many had a bustling, northern flavor. Seaman A. Knapp termed Lake Charles "essentially a northern city, wide awake, progressive and modern."

Himself an embodiment and mover of change in the area, Knapp was a New Yorker who had spent much of his life in the Midwest before moving to the busy town on the Calcasieu River in the 1870s as assistant manager of the North American Land and Timber Company. The company, purchaser of 1.5 million acres in southwest Louisiana, deliberately sought in rice a new commercial crop, and promoted the immigration of midwesterners to grow it. Knapp himself lived to transcend his work as businessman and promoter. In his old age he developed for the USDA the demonstration method of teaching improved farm methods. A major influence in the reform of regional agriculture, the method after Knapp's death was written into law, and through the Agricultural Extension Program was applied to the nation as a whole.[13]

In Texas one of the oldest of southern occupations underwent a brief spectacular burgeoning in the trailing of longhorn cattle. One result was to spread a tick-borne fever that was common in southern Texas. Later studies of the illness provided a footnote to the general history of medicine, for the demonstration that the protozoan parasite which caused the disease was spread by a blood-sucking insect helped to direct researchers to mosquitoes as a possible vector of malaria. But this contribution was of small help to the Southern cattle industry, which was struggling to reform against great odds. From long exposure, local cattle were highly resistant to the fever, but spread it among northern cattle, leading to quarantines and hostile legislation. Such problems, plus the inferior quality of the meat, contributed to the rapid decline of the booming industry after 1885.

Disease-bearing ticks were common throughout the lower South well into the twentieth century, and became a fundamental factor in preventing the development of a southern industry in high-grade cattle and dairying. Not only did the disease limit the development of markets outside the South, it effectively stopped efforts to improve the breed by importing stock. Repeated efforts to encourage a livestock boom in order to break the tyranny of King Cotton ran afoul of the tenacious fever. Even after the Bureau of Animal Husbandry of the USDA had identified the tick as the vector of Texas fever, control proved difficult; vaccines were unsatisfactory, and the ticks proved "better able to resist the effect of dips than were the cattle." Meanwhile 60 percent or more of cattle brought in from the North died.[14]

Thus, despite all efforts at reform and a multitude of local exceptions, the Cotton Kingdom continued to grow. Much of its postwar expansion came in the southern alluvial valley and east Texas. Still largely forested at the end of the Civil War, the region was turned into farmland as rapidly as the cutting of the trees permitted. Much of this southwestern area emerged as a prospering cotton bowl exploiting little-used land to bring in bumper crops that created islands of prosperity amid general regional poverty. The Yazoo delta was an example. As levees rose to protect it, the region became highly specialized and productive, with tenant-worked plantations returning high profits.[15] The production of the new lands probably helped to depress the market for cotton, making survival for those who farmed less favored regions more difficult.

It was in Texas, the newest major province of the Cotton Kingdom, that a pest hitherto little known was reported in 1894. The grayish insect, about a quarter of an inch long, punctured the buds and bolls of cotton to lay its eggs. The larvae were particularly destructive, devouring the interior of their nurseries. The creature had probably been in Texas for about ten years. After American entomologists failed, a Parisian at length identified the insect as the Mexican boll weevil, *Anthonomus grandis*.

Like any region heavily dependent on a single plant, the South had been running considerable risks. Diversification, for a multitude of reasons, environmental, economic, and social, was more than a merely abstract good. Among those reasons was the principle that the simplest ecological systems are the easiest to destabilize, since they lack the multiple checks of more complex ones. In this case, the appearance of a single pest was to cause turmoil in southern agriculture. Social factors also worked for the weevil. Phrased in the traditional rhetoric of struggle, *Anthonomus* found its opponents lacking proper weapons, devoid of discipline, and divided by arbitrary jurisdictional lines to which the small attacker was quite indifferent.

Early efforts to destroy the weevil included the use of insecticides such as Paris green, but government entomologists also urged changes in farming practice, such as burning the plants after harvest. Such social discipline proved easier to propose than enforce, and for similar reasons efforts to quarantine the infected areas failed. Farmers would neither apply the recommended measures voluntarily nor permit laws to be passed compelling them to do so. Though briefly restrained by droughts, the weevil proceeded to march eastward, greatly expanding its range in 1900–1901. By 1905 aggregate damage was apparently about $50 million. The insect spread to the north as well as the east, though more slowly, for its habits of breeding and wintering linked its life to temperature gradients. Yet it displayed great adaptability, earning to endure cooler temperatures and resisting most insecticides.[16] The last straw for some southern farmers already confronting the declining cotton prices of the 1890s, the weevil added sanctions to the demand for diversification. Unfortunately, it did nothing to change the fact that much of the southern economy hinged on the yearly cotton crop. The rural South had acquired still another parasite, but not yet an impetus to lasting change.

In many ways southern forests formed the centerpiece of the region's environmental history during the Gilded Age. From reservation to exploitation to the beginnings of scientific forestry in the United States, the woodlands exhibited the upheavals of a tumultuous time.

After the Civil War, federal policy took the South as a guinea pig for experiments in many types of reform. Land policy was one. Under the Southern Homestead Act (1866), entry was restricted and loyal settlers favored; the land unit was eighty acres and no distinction of race or color was supposed to be made in the transfer of public land. This effort to give ownership to freedmen and to revise a society which many Republicans perceived as the preserve of nabob and serf fared badly. The act was poorly administered, and the freedmen were without capital to establish themselves as independent farmers. Beneficiaries of the act included lumbermen who used their employees to file claims to public land, then stripped off the timber and abandoned the claims. In 1877 the land commissioner asked ironically, "If valuable pine lands are to be given away and the timber to be destroyed, would it not be better to enact some laws where the title can pass without perjury?"

Many thought so. Taking advantage of the confusion between the two aspects of the law—as punitive legislation, and as an endeavor to aid small farmers—the Redeemers celebrated the end of Reconstruction by leading the fight for unrestricted entry. In Congress southerners argued that the timber on public land would be better protected from fire and theft if it were privately owned. With an eye to their growing iron industry, Alabamians argued for repeal to insure their entrepreneurs access to coal land. Despite some northern and western opposition—largely on grounds that the land at $1.25 an acre would be too cheap, or that the national Homestead Act ought to be followed, or that monopolistic timber companies would take over—the repeal passed handily, with southern legislators almost unanimous in support.[17]

Not surprisingly, large areas of timberland went to lumber companies and speculators, most of them nonsoutherners. In a hectic atmosphere reminiscent of the flush times of the 1830s, land trading boomed. Especially in Louisiana and Mississippi, there was for a time more money to be made in speculation than in logging. Five and a half million acres of federal land were sold in the five southern public land states. In 1888, the last year of the boom, private entries amounted to 1.2 million acres. This trade in land, supplemented by simultaneous efforts of the railroads and the states to dispose of part of their own holdings, largely benefited northern owners, processors, and speculators. Sixty-eight percent of the lands sold between 1881 and 1888 were acquired by northern lumbermen and dealers. In Florida during 1882 a lumberman on the lookout for choice tracts reported, "The woods are full of Michigan men bent on the same errand as myself." Special trains ran from the North. And though southern buyers accounted for a third of the federal land sold, these sales too probably represented in considerable degree northern investors using, or being used by, southern entrymen.

Alarmed by the loss of potential farmland, southern lawmakers did an about-face, and in 1889 led a campaign in Congress which resulted in the restriction of entry once again to homesteaders. In the meantime, however, a substantial transfer of title to a major regional resource had taken place. Land had shifted from public to private hands, that is, from governments over which the southern voter had some influence to private and corporate owners over whom, as yet, he had none. The fact that most of the new owners were absentees emphasized not only the South's continuing status as a colony but its practical acquiescence in that status. The loss of what ought to have been a self-renewing resource in perpetuity was a grave one. The failure of policy on the part of regional leaders was not lessened by the fact that, in the end, some of the new owners proved better stewards of the land than southerners had been. The forest was still seen as a commodity, not as a functioning part of the landscape; the fact was missed that trees were both a valuable crop and an essential protection to the soil. Hardwood forests in the Piedmont helped to hold the highly erosive soil; regional warmth and moisture enabled cut trees to be quickly replaced; fast-growing pines, with taproots to search out deep nutrient sources, were exceptionally well-adapted to the poorer soils. No alternative to the alienation of the forests was proposed except to turn them into agricultural land which, without extensive reforms, would have meant an environmental misfortune of another sort.[18]

Meanwhile the national lumber industry had moved in. This was not a development *de novo:* the 1850s had seen a great increase in southern lumbering operations. In the Gilded Age, however, the region had for the industry something of the character of a new frontier. In 1865 the South possessed great uncut hardwood forests and about twice as much pine timber as the rest of the country combined. The main problem, as with most frontiers, was transport. Because of the war, the forests were probably less accessible than they had been five years earlier. Repair and construction of rail lines went apace during the years that the federal government was attempting to restrict entry; during the same years, northeastern forests were being rapidly depleted by an industry which was now steam-driven. In succeeding decades the center of the lumbering industry shifted southward, into Appalachian forests, Louisiana cypress stands, and pinelands which could now be cheaply bought up. Sawmills multiplied, changing much of the life of the backwoods, paying cash wages to rural people, and drawing black and white farmers to the lumber camps. Reforestation was not attempted, nor were seedlings protected from fire and cattle.[19]

Logging in the southern Appalachians reached its peak in the years between 1880 and 1909. Still largely dependent on rivers to move the logs, the operations were wasteful in the extreme. Cut timber floating downstream was left to rot when it became caught on banks or bends. In Kentucky, timber floated down to Louisville, Nashville, Frankfort, or Cincinnati, but by the opening of the twentieth century railroads were enabling the mills to move above the "head of navigation" for rafting. In Arkansas the slow revival of lumbering on the Ozark highlands after the war brought timber rafts down the Saline and Ouachita to mills at Monroe, Louisiana. The 1890s saw the growth of large-scale lumbering operations, as northern companies sought new sources of wealth west of the Mississippi. In Louisiana, Lake Charles had become a center of production and marketing after the Civil War. At first, mills drew their raw material from independent loggers, but with the arrival of northern lumbermen in the 1880s they shifted to mill-owned timberlands because large-scale operations demanded a constant supply. Here too the railroads extended the scope of logging; the Southern Pacific (1880) in particular opened vast new markets, ending the area's former dependence on exports by water.

The same decade brought new investments and heavy logging to southern Mississippi, long an almost virgin area. The burgeoning of the yelow pine industry, however, began only in the late 1890s. Logging under the usual motto of "cut and get out" also became widespread in East Texas. In Alabama the industry grew in tandem with naval-stores production, then expanding from its traditional locus in the Carolinas into the southern parts of the Gulf states. Lumber products remained the more valuable of the two, but both endeavors did damage far beyond any practical need. In 1901 plant scientist Charles Mohr complained of the many activities that were depleting the longleaf pines: primitive methods of extracting resin were killing trees needlessly; forest fires were, as ever, common; herds of domestic animals devoured seedlings. In a region beautifully adapted to the "spontaneous reproduction" of the trees, these forces, he believed, would soon exterminate the longleafs. As he wrote, the pine barrens, especially in Georgia, Alabama, and Mississippi, had become the most important naval-stores producing region in the world, and exploitation was everywhere "swift, ruthless, and wasteful."[20]

The rise of the Alabama iron industry was assisted by the acquisition of cheap public land covered with forests and underlain by coal. Though the area's first blast furnace predated the state of Alabama itself, the full revelation of local mineral wealth was the work of state geologists in the 1840s and 1850s. Damaged by Union raiders during the Civil War, the industry benefited during the same period from Confederate subsidies, and afterward from the aid, capital, and expertise of northern investors and entrepreneurs. Birmingham, founded in 1871, boomed during the 1880s, and by 1890 the state was turning out 10 percent of the nation's iron. Ironmasters, reported a land commissioner, "furnished money to their employees, many of them ignorant and lawless men, to enter the lands in the vicinity of the furnaces and mills for the sole purpose of acquiring the timber thereon." Ore deposits were also gained through the use of dummy entrymen.[21] For all the fraud and waste involved, this was a permanent addition to the regional economy, threatening enough to northern steelmasters to provoke a flurry of discriminatory measures designed to prevent the further growth of a serious competitor.

Yet from any point of view except that of immediate personal profit for a few, the South's treatment of its forests in the Gilded Age was fundamentally in error. Every practical advantage actually secured might have been gained at a slower pace by governments less hasty to alienate their patrimony. The flush times of the 1880s maximized loss in a resource easily renewable, and this from any point of view, commercial, environmental, or esthetic. Not only were the best trees cut, the worst were left to reproduce. Destruction did not stop with the forest. The relationship between forests and soil, rivers, and wildlife amplified the losses, implying disruption of the linked systems which constituted the natural regimen of the landscape.[22]

In view of events in agriculture and lumbering, the rapid decline of southern wildlife was not surprising. In the course of a generation or two, beginning in 1860, the first major regional extinctions since the late Pleistocene took place. For this there were three basic causes: the loss of habitat, regional conservatism and poverty, and the progress of the age. Hunting for food continued, probably urged on by a declining diet. Meanwhile the invention of the refrigerator car and the demands of the international fashion industry opened new and profitable markets to professional hunters with effects that were nowhere more keenly felt than in the South.

The most remarkable and best-known extinction was that of the passenger pigeon, primarily a northern phenomenon but a major change for the South, which had hosted the great annual migrations for uncounted centuries. Like many other species, the pigeon fell victim to the growth of market hunting made possible by the railroads. The decade that doomed the creature was apparently 1871–1880. Though its rapid disappearance may have resulted partly from infection with trichomoniasis from domestic pigeons, the root cause was destruction of habitat combined with hunting in the northern nesting grounds. Hunters took their usual advantage of the bird's spectacular gregariousness. The flocks had always drawn enemies and demanded an immense food supply; the pigeon had survived a multitude of enemies by sheer weight of numbers. In the age of the breech-loading shotgun and the refrigerator car, which entirely separated the act of killing from any consideration of personal hunger or need, this defense was no longer good enough, especially after some centuries of increasingly severe human attack. The last individual pigeons died in captivity early in the twentieth century, leaving as their memorial a lesson in man's ability to exhaust the seemingly inexhaustible.[23]

A lengthy list of other American species either declined radically or disappeared altogether during the same period—the eastern elk, cougar, and timber wolf, the red or Florida wolf, the Oregon bison, the Badlands bighorn sheep, the Labrador duck, the Carolina parakeet, the whooping crane, and the ivory-billed woodpecker, to name only a few of the more famous. The destruction was much more than a regional phemomenon. These were, after all, the times when New Yorkers ate game shot in Minnesota, and when processing centers in Kansas City and Chicago received every week trainloads of "ducks, geese, cranes, plovers, and prairie chickens"—a continental spoil.[24] Game, song, and insectivorous birds alike hung in the markets of Raleigh, Atlanta, Norfolk, New Orleans, and other southern towns and cities. In the apparent race to destroy a national resource, to say nothing of a dower of interest and beauty, the South was not unique; neither was it a laggard.

Yet state efforts to preserve favored species had never entirely died out, and by the Gilded Age had begun to show signs of revival. For evident economic reasons, fish had long been a favorite subject of legislation, and the effort led by Spencer F. Baird of the Smithsonian to protect and revive the fisheries gave new currency to an old and practical form of conservation. The laws of the southern states during the Gilded Age sought to prevent the use of particularly destructive methods of fishing, to safeguard major seafood industries, and to provide some protection to migratory species. In 1873 Kentucky tried to regulate the means of catching fish above the falls of the Cumberland, and in a series of later laws forbade obstructions which might hinder the passage of fish, as well as the use of dynamite or poison. Penalties included fines and the workhouse. In 1876 Georgia imposed a "closed time for shad" over weekends, and forbade the seining or netting of mountain trout. Arkansas in 1879 denied the more ferocious of its Izaak Waltons the use of any "intoxicating or stupefying liquid" to catch fish, added the use of dynamite to the proscribed list in 1893, forbade seines and gill nets in 1897, and in 1903 declared—a position which the U.S. Supreme Court upheld—that all fish not in private ponds were property of the state. Louisiana, a coastal state with much to preserve and much to lose, in 1870 imposed an April-to-September closed season on its oysters, and ten years later sought to halt the poisoning of fish and to regulate the use of traps and dams for fishing. In 1881 Florida followed the fashion by forbidding gill nets and small-fish seines, and the catching of shad between April and December. Vitiated by lack of systematic enforcement, such laws at any rate pointed to one of the most pragmatic sources of conservation—the need to preserve, and hence the necessity to treat as a managed resource, a fundamental source of food.

In the regulation of hunting, more complex motivations appear to have been at work. Birds followed fish as objects of concern. In the 1870s the tradition of shotgun ornithology was still strong, John Burroughs, the naturalist, wrote in *Wake-Robin* (1871, 1877), "First, find your bird; observe its way, its song, its calls, its flight, its haunts; then shoot it (not ogle it with a glass), and compare it with Audubon." Change in this cheery viewpoint was coming, however. Establishment of the American Ornithologists' Union (AOU) in 1883 brought into being a national organization concerned with preservation. Sportsmen worried about declining gamebirds, and many farmers proved to be educable on the role of insectivorous birds. Casual observation, indeed, had already provided a strong argument in favor of insectivorous species, and state laws protecting such birds sometimes appeared on the books before proof of their value to agriculture had been adduced. Insect damage was increasing throughout the United States, as cropland spread and the transport revolution carried pests to crops as well as crops to market. Federal backing for research on birds was voted, and the Biological Survey established. The AOU devised, and with the Audubon Society's aid pressed the states to adopt, a model law enumerating game birds and protecting the nongame.[25]

Early state laws were sometimes precocious; all, it would seem, were ineffective. In 1877 Louisiana made it illegal to catch, kill, or "pursue with intent" any whippoorwill, sparrow, finch, oriole, bluebird, swallow, nighthawk, or blackbird, or to rob or destroy nests or eggs. Songbirds were protected, "especially the mocking bird," unless they were trapped or netted for domestication—a remarkable exception to modern eyes. Birds feeding on crops were likewise excepted, making them fair game for irritated farmers.[26] Despite loopholes and the prevailing lack of enforcement, however, there was an evident and widespread feeling that some birds ought to be protected. Between 1861 and 1878, Kentucky, West

Virginia, and the District of Columbia protected nongame and insectivorous species. A melange of state and local laws regulating the hunting of game species also were enacted; Arkansas, for example, attempted in 1885 to regulate the hunting of turkey, grouse, and quail. Such laws had followed the march of civilization and depletion into the richest regions of the Southwest, where the last decades of the century were bringing increased complaints from hunters of a lack of game.

Depletion was, of course, still more notable in the Northeast, where what was later to be termed the conservation movement found many of its roots. There a feature of the Gilded Age was a new intensity of interest in bloodsport. The celebrated Boone and Crockett Club (1888), founded by Theodore Roosevelt, George Bird Grinnell, et al., embodied the aims of the new sportsman's movement, its moralism, *macho* ideal, and political orientation. Driven often by potent compulsions to prove there was more leatherstocking than silk stocking in their makeup, such men led the way to a redefinition of the hunt. To them hunting was not, save in rare cases of personal injury or peril, a search for subsistence; it was never a matter of grasping for a salable commodity. Rather it was the pursuit of manly recreation amid natural scenes. Though big game was their preoccupation, they made protection of many forms of wildlife part of their agenda.[27]

While new leadership gathered in the North, another wave of market hunting began, taking a form particularly offensive to the puritan spirit. Slaughter for food might be justified, however speciously; slaughter for adornment was more difficult to defend. In the early 1880s Parisian milliners launched a generation-long fashion for feather-decorated hats. Hunters at once spotted a new source of profit. Inedible water birds were now prey. Well-armed men invaded the great rookeries of the southern swamps. They killed nesting egrets, herons, and ibises, leaving the bodies to rot and nestlings to die without parents. On a trip south of Tarpon Springs, Florida, in April 1886, naturalist W.E.D. Scott found that market hunting had reduced the breeding herons near Anclote Keys from "literally thousands" to about a dozen. A former rookery near Clearwater Harbor held none. He came upon a newly slaughtered colony of some 200 reddish egrets, American egrets, Louisiana herons, and little blue herons. "It would be difficult," he wrote, "for me to find words adequate to express, not only my amazement, but also the increasing horror that grew upon me day after day as I sailed southward. . . . The great Maximo rookery at the mouth of Tampa Bay . . . was a deserted mangrove island." But it was not only the spectacular semitropical species that suffered; hundreds of miles to the north, hunters also devastated the colonies of terns and gulls that bred on Cobb's Island, Virginia, north of Cape Charles.[28]

The South, of course, was not uniquely victimized. Milliners levied tribute on the world, and New Guinea, India, Africa, and South America were their major sources of supply. The South was peripheral to the main tropical sources of supply, but unfortunately for its birds, close to one source of demand, the northeastern cities. The feather trade brought grave danger to many species hitherto safe, and a new boost for conservation.

State laws preceded the assault of the plume-hunters, and increased in response to their depredations, but appear to have been everywhere ineffective. In 1877 Florida forbade the destruction of eggs, nests, or young of any seabird or bird of plume. Two years later it forbade killing any birds for their plumes. In 1891, by overwhelming margins, the Texas legislature protected shorebirds and their eggs. But the states, even when they acted, thrust enforcement on the counties, and a patchwork of county ordinances supplemented the state laws. No special police were set up, and by all accounts sheriffs were not normally active in enforcing game laws of any sort. Guns were common and, especially in the rural South, the pay for feathers looked excellent.[29]

The importance of state laws, however, grew with new enactments at the national level. Agitation in Congress, led by the National Game, Bird, and Fish Protection Association, the League of American Sportsmen, and other groups resulted after many disappointments in the enactment on May 25, 1900, of a bill drawn up by Congressman John F. Lacey of Iowa. Twenty-nine southern representatives supported the bill and eighteen opposed it. By this law, the interstate shipment of game killed in violation of state law was made a federal offense. This archetypal act of federal-state cooperation struck market hunting at its weakest point, the need to transport its victims, their meat or plumage, to the settled parts of the country. The Lacey Act gave meaning and force to many a hitherto ineffective state law and inhibited a traffic which neither state nor federal law, nor both in tandem, were able to choke off at the source.

Many aspects of Gilded Age society—its excesses, its social changes, its gift for discovering new modes of organization—had helped to father the conservation movement. While the movement was national, both southern birds of plume and western big game were important objects of concern and important symbols to a public which required images with star quality. As the country turned the corner into the twentieth century, the struggle over wildlife signalled major changes in the way Americans lived, in their political organizations, and in their relationship to the environment and the dwindling resources of the land.

Forestry too received the early attention of scientific conservationists.[30] From the narrowest commercial point of view, use without replacement was in the long run an absurdity. From the angle of sportsmen, what was being destroyed was the context of their pleasures. Soon to be enunciated by the woodsman John Muir was a third view in which the forest was neither a commodity nor a resource nor a playground but something akin to a cathedral. All were to play a role in the national conservation movement. Effective action to achieve long-term use under the guidance of utilitarian science was the work of specialists and bureaucrats, businessmen and politicians.

The initial impulse came from the American Association for the Advancement of Science which in 1873 set up a committee to promote legislation before Congress and the states to protect forests and encourage the cultivation of timber. Over the next decade, as southerners worked hard to strip Homestead Act protection from their woodlands, the nascent movement toward protection grew in the North. The sequence suggests another side to the familiar image of the Gilded Age as a great national barbecue to which the South came late. The North, earlier at the feast, also contemplated sooner the possibility of a table littered with dirty dishes and polished bones.

Except for naval purposes, forest protection in America had no such lengthy history as game protection. The resource was of such evident importance, however, that once convinced of a serious threat the nation's leaders moved with unwonted speed. Early efforts by Interior Secretary Carl Schurz drew attention to the problem. The American Forestry Association was organized in 1875; the first American Forest Congress met in 1882 with several governors in attendance. Within the government, Land Commissioner N.C. McFarland had urged in 1864 the creation of forest Preserves. In 1876 Congress provided for a forestry agent in the Department of Agriculture—an early and enduring bureaucratic embodiment of the view that trees are a crop. In 1885 New York State set up a commission on forests; president Grover Cleveland, a product of the Empire State, urged legislation to protect forest resources; his secretary of the interior, Mississippian Lucius Q.C. Lamar, joined the president and Land Commissioner William A.J. Sparks in renewing the call for forest reserves. Authorizing legislation passed in March 1891 with an assist from the Boone and Crockett Club, and during the Harrison and second Cleveland administrations, reserves of some thirty-three million acres were set aside in western states.[31]

Meanwhile, the scientific study of forestry had begun in this country at Yale University (in 1873), and by the end of the 1880s was pursued in a lengthy roster of institutions, including the University of North Carolina.[32] In this state, so long a major producer of forest products, took place by happenstance the first actual practice of scientific forestry in the nation. On George W. Vanderbilt's Biltmore estate, the youthful Gifford Pinchot, a forester professionally trained in Europe, took on the management of the ill-used 7,000-acre forest during the 1890s. Here too his successor, Carl.A. Schenck, opened the Biltmore Forest School in 1898. Meanwhile Pinchot moved on to take over the Division of Forestry at the

Department of Agriculture, to share his enthusiasms with Theodore Roosevelt, and to impart to the future president and ultimately to his administration his own understanding of forest management as the scientific exploitation of a renewable resource.

Yet the brief presence of this gifted outsider in a most atypical part of the forest (Biltmore would have been unusual anywhere, and in North Carolina was an Arabian Nights apparition) was not the whole story for the state, or the region. As head of the North Carolina geological survey, Joseph A. Holmes, a professor at the state university, proposed creation of a national forest in the southern Appalachians and published reports on the state's resources which urged the Pinchotian doctrine of efficient use.[33] By this time many in the South were following the national trend. Southern legislators had led the way in 1889 to reimposing the Homestead Act restrictions on land entry. A study of timber resources by state geological surveys was voted by Alabama in 1896. At the same time, some Mississippi valley congressmen were being drawn into the movement because they believed that forests retarded runoff and so helped to control floods—an important link between forest conservation and the preoccupying environmental concern of the valley.[34]

Progress, however, came up against heavy and continuing opposition fueled by the quick profits of heedless lumbering and the force of custom. Arbor Day ceremonials and tree-planting legislation were one thing; no one objected to trees being planted. Efforts to protect and manage standing forests were quite another matter. The transfer of forests from public to private hands continued to characterize the age. In 1870, according to one estimate, three-fourths of the nation's standing timber was publicly owned, while in 1911 about four-fifths was privately owned.[35] The trend toward larger organizations was the most important offsetting factor of the same generation. Whether public or private, long-lived and wealthy organizations tended toward the longer view; great enterprises were far more apt to hire foresters and steer toward the ultimate returns of systematic management. As the economy became more complex, impersonal, and artificial, it became better able to conserve as well as to exploit its resources.

In the alluvial valley of the Mississippi, the movement toward reclamation and flood control developed during the Gilded Age without reference to conservation of any sort, but rather as a necessary solution to an inescapable local problem. At a later time, by a natural convergence of aims and rhetoric, river basin development (including both leveeing and drainage) merged into the liberal tradition of natural-resource management. When it did, it brought with it a tradition of federal action, contributed in great measure by southerners, that was portentous for all American waterways.

Not that the South was first at the rivers-and-harbors table. When such expenditures soared after the Civil War, most of the money flowed to the states that had rallied to the side of the Union. It was not until the mid 1870s that southerners found themselves in a condition to bid seriously for the federal dollar. The influx of Whiggish conservatives from Redeemer governments in the Mississippi valley brought to Congress a number of men who had strong personal and political stakes in flood control. Though possessing impeccable Rebel credentials, they felt little prejudice against midwesterners who faced problems similar to their own. No states were in worse need than Louisiana and Mississippi. In 1865 the Corps of Engineers found fifty-nine crevasses in the levees south of Red River. Local poverty and bad engineering remained serious burdens as the levee districts tried to repair the damages and long neglect of wartime. In the decade that followed the war, one-eighth of Louisiana's mainline levees caved into the river, a tribute to its riparian owners who wanted to enclose as much land as they could, whether stable or not, in levees built as quickly as possible.[36] Mississippi's Yazoo basin levees were in poor shape, the cottonlands flooding as regularly as the cotton and sugar country across the river. Costs closed the circle. Without profits, how could farmers afford levees? Without levees, how could they grow crops?

Making matters worse, different congressional districts up and down the river competed for federal money. Various interests—farmers, steamboaters, railroads, lumber companies—differed drastically in their needs. "The fact is," said Louisiana's Representative Randall L. Gibson, "the Mississippi was in nobody's district. . . . We were divided in our views. . . . Nothing could be done and nothing was done."[37]

Gibson, a Lafourche sugar planter before the war, afterward a New Orleans lawyer with continuing ties to the sugar interest, was also a man with fair claims to be an intellectual. Yale-educated, later to be a regent of the Smithsonian and a founder of Tulane University, he scarcely fit the stereotype of the Civil War brigadier in politics. His manner was aristocratic, his speeches low-keyed by the standards of the time, and oppressed with facts. While travelling in Europe he observed with interest the commission which had been established to control the Rhine River. In 1876 he submitted a bill to set up a commission of experts to oversee improvement on the Mississippi. Explicitly, the proposal was intended to bring the expertise of the country's growing civil engineering profession to bear upon the great river. (Gibson was friend and admirer of the St. Louis engineer James B. Eads, builder of the jetties at the mouth of the river, and later endeavored without success to win him the presidency of the commission.) Implicitly, there was quite another side to the commission. All the quarrelling factions interested in the river could unite in winning a single appropriation; afterwards they would fight as before over its allocation—but with the money in hand.[38]

At first the bill went nowhere. But Gibson was persistent, and the twined promises of a developed river—improved flood control and improved navigation—were a powerful unifying force even in a fractious Congress full of ill-feeling, after the bitter elections of 1878, between Blue and Gray. Gibson reintroduced his bill in 1878; Ohio's James R. Garfield swung behind it at a critical moment, saving it, in Gibson's opinion, from defeat; Senator Lucius Q.C. Lamar of Mississippi proposed the final compromise that passed both houses. On June 28, 1879, President Rutherford B. Hayes signed the first law that committed the federal government—a merely verbal commitment as yet—to attempt the control of floods on an American river.[39]

What followed was no grand endeavor but some years of tentative essays and much confusion. The sources of both lay about equally in law and in engineering. In fact, nobody as yet knew how to control the Mississippi, and different engineers held to dogmatic but contradictory opinions. Another problem lay in the Constitution. The Commerce Clause gave many lawmakers grounds to doubt the legality of federal spending for flood control, especially those who wanted money for navigation improvements in their own districts and who saw in the Mississippi a bottomless rat-hole for federal funds. Abstruse quarrels among the commissioners over hydraulic theory, and experimental work done by the Mississippi River Commission (MRC) engineers (army officers held three of the seven commission seats and carried out all the practical work) characterized most of the 1880s. Persistently, and with increasing effect, Gibson pushed for levees, and from 1882 on, growing sums went into the embankments. At first the Commission justified such work on grounds that levees were essential to good navigation. Attempts by some Commission members to slow down levee-building encountered congressional threats to purge them.[40]

Even under these somewhat confused conditions, a fundamental problem early began to emerge from the flood-control work. Though not easy to interpret, gauge readings appeared to indicate that flood heights on the river were rising. As early as 1884 a member called the phenomenon "a great cause of anxiety to the commission." Early engineers, both military and civilian, had predicted such a rise, which indeed was not difficult to foresee, since, if the river's floodplain was cut off by levees, it was hard to see how floodwaters could go anyplace but up. Only the most sophisticated theorists—the army's Andrew A. Humphreys and Henry L. Abbot, and the civil engineer James Eads—dissented from this common-sense view. Humphreys and Abbot held that the water would go up, but not enough to cause difficulties, since the levees could easily raised enough to contain it. Eads believed that the river would go *down* if leveed, that is, would scour out a deeper channel for itself. He wrote that "the entire Alluvial basin from Vicksburg to Cairo can be lifted, at it were, above all overflow."[41] In this case, experience tended to confirm common sense at the cost of ingenuity. The river did not deepen its channel in any consistent way; instead, floods rose beyond Humphreys's and Abbot's expectations. Up to 1890, valley dwellers had experienced the costs of meddling with nature too little. Now they would experience the reverse.

The Commission and its supporters faced the problem with increased leverage in the government. As levees were repaired, raised, and extended, a new form of federal-state cooperation came into being. State levee districts supplied most of the funds while the MRC supplied additional money (about one-third of the total expenditures would probably be a close guess), plus engineering advice, standards, and direction. As investments and population increased in the floodplain, a great national lobby took form. Railroads laid vulnerable trackage behind the levees. Northern banks and corporations invested in transport, land, and timber, counting on the levees to protect workers, equipment, housing, and so forth. Leveed areas like the Yazoo basin became major producers of cotton; the Louisiana sugar bowl depended on its walls of earth. The New York Chamber of Commerce no less than the St. Louis Merchants Exchange could be depended upon to support appropriations for the levees when needed.[42]

In 1890, floods of extraordinary magnitude overwhelmed the alluvial valley from Cairo to the Passes. As a formidable spring rise flowed south between the Tensas basin in Louisiana and the Yazoo basin in Mississippi, both of which were now protected, the MRC reported the river "higher than ever before known." Endangered by the flood was a region which had grown more populous by reason of a "great awakening of enterprise," stimulated in part by the prospect of protection. Roused by the flood, Delta lawmakers seized the opportunity to push through a new appropriation and to eliminate the old proviso that levees could be built only to protect navigation.[43]

The river continued high for several years, average maximum flood heights for 1890–1893 exceeding those of 1886–1889 by about 7.5 feet at the Red River Landing gauge. A major victim of the flood years was a flexible river policy. Hitherto experimental and inclined to try a variety of control methods, the Commission slipped rapidly into the celebrated policy of "levees only." Levees became its main work and *raison d'être*, though always supplemented by some revetment work and by dredging to open a reliable shipping channel. By the mid-1890s the Commission had little choice, for the levees already built had so far raised the floodline that levee-building, repair, and protection absorbed most of the funds available. Of course, as new levees were built and overflow areas were further cut off, the floodline rose still higher. Commission members were probably not unaware of the fact that the dynamics of their work assured them jobs in perpetuity. However that may be, levees had to be raised and raised again, and, as they were raised, strengthened; and costs escalated with every square foot of cross-section.[44]

Federal power had proved to be a formidable fact. Quite possibly, more was accomplished in improving the levee system over the two decades following creation of the Mississippi River Commission than in the century and a half between the first French embankments and 1879. Levees were not only bigger but better; work was continuous, systematic, and informed. In 1897, for the first time in history, a major flood was passed successfully to the sea between levees. Gibson's commission had done all he hoped. But the Mississippi was also a formidable fact, and many tragedies lay ahead, as the MRC struggled to avoid the fate of the Lady of Niger

> Who smiled as she rode on a Tiger.
> They returned from the ride
> With the Lady inside
> And the smile on the face of the Tiger.

The South was a backwater of the Gilded Age, despite all its bootstrap achievements in industrialization and agricultural reform. Because the nation organized to exploit more quickly and fully than it organized to conserve, the southern environment overall suffered far more than it gained from the achievements of the age. Yet no simple summary is possible of a time which in retrospect marked the beginning of a great and complex transformation for all Americans. The time of exploitation unlimited was also the time of innovation unprecedented. Tropical medicine, scientific forestry, wildlife conservation, and the engineering of rivers all implied drastic changes in environmental relations. But all represented beginnings only. Any unravelling of the fundamental southern enigma—disease and poverty and a declining resource base—would be a task for the future.

7

Conserve and Develop

AS A THEATER OF ENVIRONMENTAL DISASTERS, THE SOUTH OF 1900–1930 offered instructive dramas. The boll weevil infestation that spread continuously during the period and the great flood that occurred toward its end were cooperative ventures, jointly produced by man and nature. But growing human intervention in nature on a grand scale had the most varied practical consequences. During the same decades a bold though as yet unsuccessful attack was mounted on the region's endemic ills, and a successful one against the last appearance of the epidemic killer yellow fever. A lively conservation movement began to offer promise, if as yet little more than that, of reversing some old and basic errors in the region's treatment of resources. It is notable that in all these events, new forms of organization—introduced by business, the states, or the federal government—and new methods of cooperation among them were at work. Joint productions were the rage.

Yet much remained unaltered, despite the remarkable innovations of the time. The opening of the southwestern oil fields promised the region a true Golconda. The growth of Miami and the south Florida region indicated that in the chrysalis of the New South, the Sunbelt (or at any rate a distinctive part of it) had begun to take shape. Neither cities nor industry, however, yet posed much of an environmental problem, except in small and atypical areas. The old problems of disease and poverty and the waste of meager resources set the tone of the time, limiting the innovations of local reformers and the national government, whose influence over the region nevertheless grew ever stronger.

The advance of the boll weevil precipitated, if not a regional crisis, then a recurrent series of local crises whose effects were felt throughout the lower South.

Moving from west to east, the insect became a folk legend and a bringer of ruin. There was little logic to the pattern of which areas were hit and which were not, save the logic of the wind (weevils moved against the wind), proximity to areas already infested, and, of course, the presence of cotton. Early in the century Texas growers faced a major disaster. In Limestone and Robertson counties "nearly half the farms were abandoned and one third of the stores in the town were closed." In Louisiana, Seaman A. Knapp declared, "I saw thousands of farms lying vacant; I saw a wretched people facing starvation; I saw whole towns deserted; I saw hundreds of farmers walk up and draw government rations, which were given them [to keep them] from want."[1]

A contemporary statistical study of the impact on Texas advanced some interesting figures. Cotton acreage in Texas had risen during 1879–1889 by 80.61 percent and increased again by 77 pecent during the 1890s. After 1899 many factors conspired against the staple—the press and agricultural leaders preached diversification and gained some following; truck farming became more important and fruit-growing increased in East Texas. Rice and the oil industry came in, as well. Nature was less than kind. The Brazos flooded in 1899 and 1902, and the extraordinary hurricane of September 1900 hit cotton acreage hard. But the impact of the weevil was the fact of greatest importance to cotton growers, for the insect spread until it destroyed about half the crop. In 1904, 1.71 acres were needed to produce the cotton that a single acre had grown before the coming of *Anthonomus*.[2]

In hard-hit counties throughout the South, wholesale and retail trade fell off as the shrinking crop drew a dependent economy with it. The "advance merchant" who had met nearly all the needs of farmers in return for a lien on the cotton crop could no longer do business in the old way, if at all. The consequences were full of paradox, for the treadmill of debt peonage was broken for many only by bankruptcy and emigration. The outflow of population in some areas caused an agricultural labor shortage despite the reduction of the cotton crop. This access in bargaining power meant cash payment to some tenants who before had received supposed credit and no accounting.

In Mississippi's Yazoo delta, an effort by planters dissatisfied with black labor to bring a white peasantry from abroad failed as the immigrants found themselves rejected by the natives and decimated by the endemic diseases. Landowners then seized the opportunity presented by the boll weevil to bring in impoverished black tenants from the hard-hit Natchez district. "The country roads of Natchez were filled with negroes, wagons, and mules," wrote two observers. "The streets of the town were filled with puzzled negroes . . . [who] knew nothing except that their merchants would carry them no longer, that they could not carry themselves, and that they had been called upon to pay what they owed or surrender what they had mortgaged." Delta planters also thronged the town, settling accounts and moving the tenants to their new homes. "The net result to these counties was, on the one hand a group of fairly well satisfied city merchants, and on the other a disorganized country, stripped of labor, farming implements and stock, empty houses on tenantless land—a picture of desolation for a counterpart of which memory must return to the devastation of the Civil War."[3] The delta went on much as before, only more so.

In some places there were permanent changes. Over much of the coastal region of South Carolina and Georgia the weevil destroyed the glossy long-staple cotton which had been grown there for two centuries. Sea Island cotton required a month longer to mature than the short-staple, which gave the weevil more time to eat and multiply. "I didn't make nothing," recalled a farmer of his 1917 crop, "the boll weevil got all of it." For two years after the first outbreak no cotton at all was grown. Survival commanded a change to upland cotton and to subsistence farming. In 1930 an observer remarked upon the changes that a few years had brought to the region. Now short-staple was grown and little of that. Fatalism was rife. "The idea that the weevil is a divine dispensation and should not be disturbed is still quite prevalent." Yet outmigration had been commoner before the weevil than after, and those who stayed survived by diversification, growing sweet potatoes and peanuts, a little rice, cowpeas, melons, sugarcane. Almost every farm had a cow and calf, an ox or horse, and the inevitable pigs and chickens.[4]

The years up to about 1920 also saw the climax of the long process of erosion on the southern Piedmont. Over the course of settlement and exploitation, but especially in the generations since the Civil War, a mass of topsoil had disappeared which has been variously estimated at six cubic miles or, more vaguely, sufficient to cover "an area equal to that of Belgium."[5] When the weevil and erosion combined forces in a hilly country such as Georgia's Greene, the results could be dismal indeed. With much topsoil lost, woods cut, cotton infested, and even the rabbits dying of a tularemia outbreak, inhabitants were left with few resources except such textile mills as could be kept running, and the distilling of illegal whiskey. Between 1923 and 1930 the black population, with its heavy proportion of tenants, decreased by 43 percent, the white by 23 percent. In such rural areas, decline and abandonment mocked the general prosperity of the time. "Dead cotton and corn stalks weathered one winter and came to the next, ground frost spewed loose their dead roots, March winds blew them over, and that year the heavy rains, which always came, washed more loose dirt than ever before into the creeks and rivers of Greene."[6]

Overall, the weevil could only be termed a calamity, an irruption somewhat comparable, in a food crop, to the potato blight of the 1840s.

Yet the Cotton Kingdom survived and grew. There was no standard for the weevil's destructiveness; its impact diffused through many different localities, modified by wind currents and climate, local mores, social organization, and wealth. By reducing output, the depredations of the weevil may have raised cotton prices, encouraging continued planting of the crop. But the basic factor in the Kingdom's survival may well have been the assistance tendered by anxious governments.

Agricultural experiment stations had been set up originally in the landgrant colleges, receiving both state and federal aid, the latter through the provisions of the Hatch Act (1887). They had spread widely in the South by the end of the nineteenth century. They carried out a melange of functions, including much applied and some basic research; they tested fertilizers, provided extension services to farmers, and did lab work when the Progressive era brought in state pure food and drug laws. When the boll weevil hit, station scientists worked with USDA entomologists to find means of control.[7] Because the improvements in farming practice they had originally proposed had failed to win acceptance, scientists turned increasingly to pesticides. In 1918 the use of calcium arsenate to poison the dew that adult weevils sipped in the morning proved effective, reducing damage to a level that was tolerable to many farmers. As production rose, the South learned to accommodate its hated guest. Despite the efforts of a legion of agrarian reformers and the year-by-year advance of a still larger legion of boll weevils, the Cotton Kingdom survived the bad years and grew with the good ones.

In 1919, Mississippi, Alabama, Georgia, and South Carolina had almost 215,000 more cotton farmers than in 1899. A major emigration of southern blacks, launched by northern job opportunities during the war, received a new push during 1920–1923 as weevil depredations combined with desperately low cotton prices. By the late 1920s prices had begun to recover and the South had forty-two million acres in cotton. The crop was "more dominant in southern agriculture than it had been a century before." Indeed, cotton had at last supplanted corn as the region's most widely grown crop. In 1879 there had been some ten million more acres of corn than cotton, but fifty years later "cotton occupied about that much more cropland than corn."[8] High prices and habit pushed growers in this directions; but science made the recovery and extension of the Cotton Kingdom possible.

The South had learned to restrain but not to destroy its new parasite, which thereby became endemic and a part of the order of things. In the process a multitude of changes overtook both weevils and men. Abandoning the varied diet of its wild forebears, *Anthonomus* learned to live on cotton almost alone.[9] Because combatting the creature demanded both capital and knowledge, those farmers who could command either were naturally favored. Throughout the cotton-growing region the costs of the infestation were proportionately heaviest upon the small producer, while money and education and wide acres took an ever more commanding advantage over the narrow margins, financial and intellectual, of the poor. The work of agricultural scientists similarly favored those who were in every way best equipped to adopt their teachings. In this the weevil infestation was a special though significant case in the more general evolution of scientific farming, which slowly and sometimes with bitter irony (given the ideals of agrarian reformers) helped to squeeze out the "less educated general farmer . . . [who] was often inhospitable to 'book farming' and to agricultural colleges which seemed remote from his needs."[10] There was a sort of specialization at work here, and it worked from both ends, shaping the pest to its new ecological niche and shaping the farmer to the task of fighting it.

In a region with such a history of illness and with so many economic problems it is not surprising that the major endemic diseases also maintained themselves against even informed and systematic attack. Yet public health campaigns did enjoy some success, particularly in preventing epidemics. Yellow fever was extirpated, while cities gained new protection against diseases spread by polluted water and probably became, for the first time, healthier than the country. Of the rural districts it might at least be said that the deficiency diseases and malaria were put under siege.

The yellow fever story is well known. Walter Reed and William Crawford Gorgas were southerners who, in the conditions prevailing after the Civil War, sought medical careers in the army. The devious paths normally followed by such careers eventually brought Reed to Cuba as head of the Yellow Fever Commission, where by a series of classic experiments he and his colleagues demonstrated the transmission of the disease by the mosquito *Aedes aegypti*. The ambitions of the United States in the Caribbean led to the attempt to complete an isthmian canal. The area chosen, Darien, was hyperendemic both for yellow fever and malaria, and Gorgas was ultimately called upon to work out and enforce, in Cuba and in the Canal Zone, a systematic program aimed at interrupting their transmission. When a substantial measure of success resulted, vector control emerged at last as the practical point of attack on the great febrile diseases. Even New Orleans adopted Gorgas's methods in 1905, halting a yellow fever epidemic in progress—an event so extraordinary as to announce in truth that tired cliché, a new era. The incalculable human and economic toll levied on the South over the course of two centuries by yellow fever was ended.[11]

Attacks on the endemic diseases were admirable but less successful. Ailments which had long proven their ability to flourish in the South and maintain themselves from year to year were far more tenacious than the introduced, epidemic ills. After 1900 the endemic diseases were associated with rural areas even more than in earlier times, and in this fact was a basic reason for the difficulty of control. To oil or screen the artificial receptacles that might breed domestic *Aedes* required mainly patience and discipline, which the dense, accessible urban populace was now ready to apply. Malaria control in towns was possible for the same reason, and by 1926 had effectively ended the disease as an urban threat. To oil the innumerable bodies of water in which *Anopheles* bred in the country, down to the hoofprints of cattle filled with rain, was quite another matter. Country ways, country distances, and the deep conservatism of the rural poor were also barriers.

County governments, state boards of health, and national foundations began to combine forces against the endemic ills. In 1912–1913 the U.S. Public Health Service opened a systematic campaign against malaria in four southern states. Aided by Rockefeller money, public health workers made excellent progress in reducing morbidity. Efforts intensified during World War I, especially in the neighborhood of military bases. In 1923 the Rockefeller Foundation established a malaria research station at Leesburg, Virginia, where the chief vector, *Anopheles quadrimaculatus*, was identified and studied. Here the nation's first training center for malariologists turned out experts to fight the disease. The victory they sought was not, however, to come for three decades. In 1919 Gorgas's able onetime assistant, Henry Rose Carter of USPHS, estimated the annual loss of the southern economy to malaria at $100 million. In 1939 another qualified researcher was to calculate the annual consequence to the South of "death, disability, and unproductiveness" caused by the disease at $500 million. Uncertain as all such calculations are (how, exactly, is one to assign a dollar value to medical effects ranging from lethargy to death?), there seems little reason to believe that general progress was being made on the countryside, whatever the advances in the towns. "In the years 1923 to 1924," recalled a Rockefeller researcher, "in southern Alabama and Georgia, I surveyed many communities, colored and white, in which malaria was a dominant factor in producing ill health and economic stagnation."[12] These were relatively good years; the incidence of the disease was shortly to rise.

Yet ways of turning laboratory knowledge into practical gains at endurable cost were being devised. Outside help formed part of the answer; federal personnel and the wealth of private philanthropists were both essential to poor southern states struggling with entrenched ills. The local response was just as critical. By 1922, 163 countries in 10 southern states were cooperating with USPHS and the Rockefeller Foundation, as were more than 100 towns and cities. The whole region was far more health conscious, and not only in the cities; the paternalism of southern mill towns led to some interesting experiments in group medical care which were considerably ahead of their time.[13]

Indicative of the possibilities and problems in southern public health campaigns of the time was the attack on hookworm disease. The bizarre life and devious ways of the hookworm were first described by European researchers. The organism was discovered in Puerto Rico by an Army doctor, and identified as a separate genus by a Public Health Service officer, Dr. Charles Wardell Stiles. Stiles and a Georgia physician, H.F. Harris, independently proved its presence in the South. Eggs in the feces of victims infected the soil when privies were not used or privy vaults were improperly constructed. The larvae entered the skin of the new host, causing a local irritation called "ground itch." At this point, as a medical historian has pointed out, the parasite was quite vulnerable. All the rural poor had to do was to wear shoes. But who was to buy them shoes?[14] Once in the bloodstream, the parasite passed through the heart to the lungs, entered the bronchi and ascended the trachea. Swallowed, it traversed the stomach and settled in the intestines. Here, if in moderate numbers, the worms might be tolerated by the host, especially if he enjoyed an iron-rich diet. In great numbers, however, they might deprive him of sufficient blood to cause lethargy, dullness, and physical malformation, especially if he was ill-nourished to begin with. If they blocked the intestine, the worms could cause a painful death.

Some early converts to the anti-hookworm cause hastily concluded that the parasite's effects explained all that was peculiar, backward, and un-American about the South. A South Carolina doctor saw in vermifuges the means of making poor whites "efficient developers of the South's and of the State's resources."[15] Actual progress in the fight was, however, slow. Education and propaganda, systematic collection of stool samples, giving lessons in the proper construction of privies, setting up dispensaries and administering thymol (an effective though dangerous drug) were the hard, inglorious tasks of a public health campaign in the trenches. A Florida doctor who participated recalled that "treatment in those days really did have its hazards. . . . You had to starve [the patients] both before and after treatment. We would run into fierce opposition from parents on this account. . . . When we would cure a case of hookworm disease and instruct mothers to make their children wear shoes until the dew or rain dried off the grass, not one mother out of four would see they obeyed these rules. So three out of every four children that we cured of this disease would have it again within a year after we cured them."[16]

Though Florida opened the campaign on its own, outside help soon became available. Stiles had embarked on a crusade against the "germ of laziness" which at first encountered amusement in the North and resentment in the South. In 1908, however, after Stiles met a Rockefeller agent in New York, the magnate announced that he would contribute one million dollars to eliminate the disease. In 1909 the Rockefeller Sanitary

Commission for the Eradication of Hookworm Disease was set up to lead the fight. For the five years of its existence, the Commission worked through southern officials to overcome local suspicion and hostility.[17]

The basic problem remained the rural South itself. The advent of the twentieth century had by no means transformed its sanitary practices. Half the homes Stiles surveyed had no privies whatever, a situation which changed slowly. In 1926 a survey of 500 farm houses in Arkansas showed that 20 percent had no toilet facilities, 75 percent had outdoor privies, and only 5 percent had indoor toilets. A survey of 947 homes of white families in Texas revealed that 15 percent still had no toilet facilities. Because such conditions were wrapped in the whole context of rural living among the poor, they were slow to change. Culture, climate, and parasites remained fused into an obdurate environmental and social reality against which knowledge and devoted effort had strictly limited and temporary successes. Statistics on hookworm continued for decades to show a very high incidence in many areas. During 1910–1914 Alabama used about $56,000 in Rockefeller grants plus $4,500 from the State Board of Health and $7,863 contributed by fifty-seven counties to attack the disease. Yet in 1916, 60 percent of Alabama National Guard personnel detailed to the Mexican border were found to be infested to some degree. This might be compared with the rate of infestation turned up by investigators in the state's hyperendemic southern counties—62 percent—before the campaign began. Even in the late 1930s, a cooperative study by the Florida State Board of Health, the Rockefeller Foundation, and Vanderbilt University found adolescents aged fifteen to eighteen the worst afflicted group (44.7 percent), and the state's Panhandle the most severely affected area (49.2 percent). No doubt the number of parasites, and the number of their victims suffering from clinical manifestations of the disease, were both significantly reduced through much of the rural South by the public health campaigns. Those who struggled with this tenacious illness, however, like its remaining victims, still had a hard road to travel.[18]

Despite impressive work by both national and local forces, the picture was similarly clouded in regard to malaria. During 1900–1920 the disease maintained itself in the Southeast, eastern Texas, Oklahoma, and the Rio Grande valley—centers where it was to survive until after World War II. Malaria did undergo peaks and recessions, as its victims developed immunity in response to repeated infections. Thus 1926–1928 saw a peak in the cycle, and 1931–1932 a decline even in the hyperendemic area. A new rise then followed which reflected the impact of the Depression—displacement of the poor, inability to make necessary repairs on houses, and the general debility caused by a poorer diet. In 1933–1935 four to five thousand Americans still died annually of the disease, nearly all of them in the South and its western borderland.[19]

Nevertheless, an indispensable basis for improvement had been laid.

Understanding the mode of transmission had led to development of methods of control in Panama which had considerable success under the disciplined government, armed with adequate funds, and the comparatively narrow geographical bounds of the Canal Zone. The technique of attacking the vector of malaria by screening, spraying, and so forth, would soon be joined by new methods of prophylaxis in the artificial antimalarial drugs, and the discovery of new insecticides. Meanwhile the rural South resisted cure. To a dismaying extent, its diseases had become part of itself.[20]

During the same years, both the South and the nation were also organizing to conserve resources other than the human. The conservation movement of the early twentieth century resulted from three major developments of the preceding century and especially of the preceding generation: the depletion of resources, the growth of organization, and the development of science. Despite the region's enduring conservatism, there was much in the Progressive conservation movement that sang sweetly to southern ears. The South had learned the sport and trade of subsidy hunting during the Gilded Age, and federal initiatives clearly had much to offer in overcoming old regional ills. Beyond this was the fact that conservation defined as efficient use did not preclude but rather coincided with rational development, and development remained the South's first priority. In the improvement of agriculture, in scientific forestry, and in river basin planning, as in public health, the new ideas of the age offered great advantages to a backward, struggling section. Indeed, the failure of the South to join the West in resisting conservation was probably a fundamental reason for the movement's widespread success. And this is not altogether surprising, since the New South creed could easily accept the leadership of the Northeast, as it had been doing for a generation, in matters of development and the treatment of natural resources.

The immediate appeal of conservation was to big business, which hired foresters and other scientists, played for subsidies, and built toward a managed future. After Gifford Pinchot took over the Division of Forestry in the Department of Agriculture (reorganized as the Bureau of Forestry in 1900) he offered expert help to timber companies. The offer was eagerly seized upon by many of the nation's largest companies, among them the Kirby Lumber Company of Texas, the Northern Pacific Railroad, and the Weyerhaeuser Lumber Company. The Transfer Act of 1905, which shifted the national forests to Pinchot's control, similarly won support from Weyerhaeuser, the American Forestry Congress, and railroadmen, including James J. Hill.[21]

Under Theodore Roosevelt, federal initiatives transformed resource policy. The White House Conference of Governors (1908) led to a national inventory of resources, a proposal originally made by the engineering profession, which hoped to see "sound business principles" applied to the nation as a whole.[22] Primarily the work of the federal bureaus, the survey included predictions of exhaustion—petroleum would be used up by 1935, available coal by 2027—which proved about as reliable as most. The basic conflicts faced by the conservation movement, however, had nothing to do with abstract truths. After centuries of catch-as-catch-can, Americans were ill prepared for the long view in handling resources. Little exploiters hoped to go their usual way, and local magnates to use up resources which they felt they owned; Congress was accustomed to *ad hoc* methods which in any case were politically remunerative; federal bureaus like the Corps of Engineers had turf staked out, and viewed many of the new departures (sometimes with reason) as power grabs.

In the struggle waged by centralizing rationality and its moral outriders against things as they were, forest policy was one central issue, river basin development another. The year 1909 is often taken as the peak in American lumbering, after which depleted resources and systematic forest management began to reduce production. Under Pinchot's guidance, and—despite the public perception of decline from Rooseveltian standards under the regime of President William Howard Taft—afterward as well, federal forest reserves grew steadily. Even more important to the nation (particularly the South), the practical view of timber as a resource and the tree as a staple took hold.

"Cut and get out" logging endangered scenery in the sloping Appalachians, leading in 1899 to formation in the resort town of Asheville, North Carolina, of an Appalachian National Park Association backed by the local Board of Trade. The idea of bringing the national park, hitherto confined to the West, to the East, where the federal government would have to buy land rather than merely reserve existing holdings, meant major changes in traditional practice. Federal studies supported, not a park, but a forest reserve. The movement grew as chambers of commerce, local politicians, the regional congressional delegation, and even lumber companies joined in. The hope by 1901 was to save the local logging industry by achieving sustained yield through scientific forestry. The Southern Railroad supplied funds; six southern states passed enabling legislation; and national conservation leaders, including Theodore Roosevelt, joined in. After an epic siege, a campaign orchestrated by the American Forestry Association won congressional action at last in 1911, utilizing the well-established federal power over navigable waterways through the claim that acquisition of watershed was necessary to protect streamflow. An impressive piece of intersectional cooperation was also involved, for New Englanders had been seeking a federal forest reserve in the White Mountains, and in the last phase of the battle Blue and Gray combined forces.[23]

The result was an important new departure for federal forest policy, and something of a bonanza for the locals. The government was authorized to buy land at the rate of $2 million per year until 1915 in the Appalachians and the White Mountains of New England. Jobs were created in roadbuilding, firefighting, logging, and the expansion of tourism. New legislation in 1913 and 1924 broadened the scope of the Weeks Act, permitting land acquisition for timber and power production. In 1940 the reservation became the Great Smokies National Park.[24] The eastern national parks and forests bore the hallmarks of good policy, not only because they brought science to bear on a major resource, but because they were popular, and profitable to almost everyone concerned. On such practical successes the conservation movement established itself as an enduring part of the national scene.

State action was generally less successful. Poverty disinclined southern states to buy forest land. In 1916, however, a group led by North Carolinian foresters and Louisiana's progressive lumberman Henry E. Hardtner established the Southern Forestry Congress and launched a campaign of public education which showed some positive results. Hardtner was an original, a businessman who had begun experimenting with reforestation in order to secure his source of raw material. In the early years of the century he achieved sustained yield. A politician, and well connected with several governors, he gave the state's forestry a bent both progressive and thoroughly pragmatic. As chairman of the Louisiana Commission for the Conservation of Natural Resources, he secured from the 1910 legislature tax breaks for reforestation. By the second decade of the century the perception of the need for scientific forestry was widespread in the South. Virginia and Alabama had set up state forestry programs. Nine southern schools had courses in the subject, the best of which was established under a grant by George Foster Peabody at the University of Georgia in 1906. In 1914 the Georgia State Forest School was set up. In Tennessee a forestry association dated from 1901; a state forester was appointed in 1914, and a bureau was established by the state government in 1921.

The pine forests, always something of a stepchild of the conservation movement, benefited from research supported by the United States Forest Service. In 1901 Charles Holmes Herty introduced a new method of extracting resin from the trees by cup and gutter. Based on French methods, the process was far less destructive to trees than "boxing", and because it was more productive as well, was widely adopted. Austin Cary, a New Englander with the Forest Service, was influential in pushing a sustained-yield philosophy which he combined with a strong bias toward private ownership and belief in the overriding importance of economic motivations. The changes that resulted from such preaching and example were not in the exploitation of woodland—which continued and indeed intensified—but in the transformation of the human perception of the forest from commodity to renewable resource. Much was implied in the change, including the recognition of an ultimate limitation on natural abundance (however erroneous the specific predictions of scarcity might be), and a willingness on the part of businessmen to take responsibility for future generations, or at any rate for future dividends. In the transformation begun during the Progressive decades, great organizations and small artifices both had a place.[25]

The wild creatures supported by forest and marsh also received attention. The national Audubon Society orchestrated much of the movement for reform, aiming particularly at coastal states where federal law had impeded but not stopped the killing of seabirds. As with the Boone and Crockett Club, private and voluntary association provided the first key to wildlife conservation. Thomas Gilbert Pearson, a Florida-raised citizen of North Carolina, became a local Audubon Society leader, impelled by the conditions in his adopted state at the turn of the century. At that time both male and female deer could be shot year-round in twenty-three of the ninety-seven counties; elsewhere there was a three-month season. There was no protection for quail in eight counties, for wild turkeys in thirteen; and no protection whatever for shorebirds, except in one county. Songbirds might be shot, netted, or trapped at any time of the year. Game hung in the markets of Raleigh, Greensboro (where Pearson taught at the women's college), Winston, and Charlotte. There were no game wardens and never had been any.

At one time an ardent hunter and egg collector, Pearson knew well both birds and hunters. He disliked sentimental propaganda, and judged that the Audubon Society itself was not guiltless in this respect. He noted that "gentlemen hunters and their families were my main supporters . . . and wanted market-shooting and plume hunting brought to an end; while men who did not hunt seemed to take no interest in laws to protect wild life." Here as elsewhere, the split between sportsman and market hunter over the purpose and etiquette of the chase first made conservation of wildlife possible. The villains of the piece were, in the main, unlettered folk who talked with "simple frankness" of their depredations, and viewed the killing of shorebirds as they did the catching of fish—in each case they were "farming the sea for its products." At least one plume-hunter endorsed the need for laws to protect his prey "until they can catch up to their numbers again," since otherwise the "plume-bird business will sure come to an end."

With the energy of a man who had found a profession and a cause in one, Pearson harangued the General Assembly of North Carolina in 1903 and obtained a law, not only protecting nongame birds, but establishing seasons and a nonresident license fee of $10 for hunting game species, with the proceeds turned over to the Audubon Society to hire wardens. This surprising use of a private organization to carry out police functions reflected the poverty of a state which was still congratulating itself on being able at last to provide a four months' school term in every county. With earmarked funds drawn from outsiders, Pearson hoped to provide enforcement free from politics; but there were many disappointments. Hunters counterattacked, the legislature withdrew half the state's counties from the Society's jurisdiction, and Pearson and his cohorts were obliged to mount a sixteen-year campaign to establish a state agency to enforce statewide game laws.

Meanwhile the Society fought for protection laws in most of the Atlantic and Gulf states. Coalitions varying from state to state secured model laws from Arkansas in 1897, Florida in 1901, Kentucky in 1902, Louisiana and Mississippi in 1904, and Alabama and West Virginia in 1907. As in the Carolinas the Audubon Society took on the burden of enforcement. Always, license fees, especially for state residents, were a sticking point. As for Pearson himself, he shifted his base of operations in 1911 to New York, where, in the words of a leading American ornithologist, he became so closely associated with all the Society's doings that "the biography of the one is virtually the history of the other."[26]

Each southern state followed its own pattern of conservation progress, keyed to local politics and personalities. Louisiana was not only one of the two most important states in the region for wildlife, but a leader in a field in which it had done little until late in the nineteenth century. A state Audubon Society was set up in 1902, and its campaign of public education astutely stressed the economic importance of Louisiana's birds. Indeed, to a state so beset by voracious insects—notably the tobacco worm, cotton borer, and corn borer, which were said to be undergoing a population explosion because of the destruction of insectivorous birds—and with the boll weevil moving east, feathered allies were not to be disdained. In 1904 an excellent law protecting nongame birds was passed, and in 1908 the state set up the first conservation commission to be established by a legislature, the lawmakers citing in the preamble to the act the recommendations of Roosevelt's Council of Governors. State wardens were provided to enforce the new game laws.[27]

In 1903 Virginia passed a model law to protect its seabird colonies, but enforcement lagged and for a time the state seemed to be moving backward; legislators showed a common southern reluctance to require hunting licenses for residents, and because of the resistance that developed, a state Commission of Fisheries and Game was not established until 1916. In South Carolina the Audubon Society at first attempted, under state authority, to function as a game commission. In 1909 it recommended a state commission instead, funded by license fees. With the eastward progress of the boll weevil making bird protection more attractive to farmers, a state game warden system was created in 1911. In this state, too, however, the resident license fee proved a sticking point, and in consequence the program was not adequately funded.

By contrast, Alabama provided a law in 1907 which was one of the best in the country, containing the American Ornithologists' Union model and additionally providing for licenses and a state game commission. In 1911 Georgia enacted a law modeled on its western neighbor's, giving teeth to the law it had originally enacted in 1903. Enforcement as usual proved difficult, but the Georgians undertook to replenish fish and game stocks. Among the coastal states, Mississippi and Florida were laggards, yet the former passed a nonresident license law in 1904, and prohibited the sale and export of protected species in 1906. Florida was swift in theory, slow in practice, for though it passed the model law in 1901, it provided no system of state wardens until 1925. Among the inland states, Tennessee passed a model law in 1903, a nonresident license law, and appointed an unpaid game warden.[28] Many of the states were active, and they followed, very often, the best guides—the National Conference of Governors, to which the Louisiana law paid tribute, the Audubon Society and the Russell Sage Foundation, and the more progressive northern states. Alabama's law, for example, was modeled on Pennsylvania's.

Slow progress took place against continuing difficulties. In the first decade of the century the hiring of wardens was still being done by the AOU and the National Association of Audubon Societies. Later, in 1909, some federal wardens were provided by law. However they drew their salaries, such men faced a thankless and dangerous task. Many were threatened, some assaulted, and celebrated murders of wardens by plume hunters occurred on Oyster Key in 1905 and at Charlotte Harbor, Florida, three years later. Pressures against the trade, moreover, helped to raise prices for feathers and so to provoke new hunting. The federal government moved to help, in 1900 seizing 2,600 gulls in a Baltimore millinery establishment which had been shipped in defiance of Maryland law from Morgan City, Louisiana. The first federal bird refuge was set up by President Theodore Roosevelt at Pelican Island, near Orlando, Florida, in March 1903. Later he set up eight more reserves in Florida and four in Louisiana. A remarkable private effort at preservation was the creation by Louisianian E.A. McIlhenny on Avery Island, a hill-like formation over a salt dome in table-flat South Louisiana, of a refuge for snowy egrets and herons. Established in 1892, the refuge by 1902 contained 1,000 pairs of nesting egrets, and in 1909 McIlhenny was able to ship 3,000 birds to south Florida to aid in restocking an area devastated by plume hunters. McIlhenny purchased thousands of acres of marshland as a refuge and persuaded Mrs. Russell Sage and the Rockefeller Foundation also to buy vast acreages for the same purposes.

Despite such endeavors, a recent historian has concluded that what ultimately destroyed the plume trade was neither private generosity, nor the courage of wardens, nor refuges, nor laws. Fashion had created the trade when the Victorian middle class, in its vast numbers and new-found respectability, undertook to ape the ancient aristocratic fashion of attiring women in the plumes of birds. The change in the status of women brought in by World War I altered fashion once again. "Billowing trimmings" disappeared. Clothing became simpler and more functional. In the home or out of it, middle- and upper-class women had many things to do other than turning themselves, or imitating those who turned themselves, into bizarre and sumptuous *objets d'art*. The decline of demand crushed the trade. Meantime the furor over the birds, much of it undoubtedly provoked not by practical considerations but by the same beauty that made the feathers a milliner's delight, had helped to bring a new era not only to the nation's treatment of birdlife, but by extension to its treatment of many other forms of wildlife, as well.[29]

The basic regional need was development, keyed often to the discovery of new resources and the invasion of new regions. Within the New South, enclaves of nascent Sunbelt began to take shape, and in all states local boosters earnestly imitated the Yankee norm.

The rapid growth of Miami during a real-estate boom in the mid-1920s helped to signal a major change in the life of Florida and to provide a portent of things to come. Drainage of the Everglades, a perennial project of boomers and scientific optimists alike, had advanced early in the century under a reforming governor. Later, when the work had to be financed by land sales to speculators, it became a persistent source of fraud. By the 1920s, however, water levels in the Everglades had been significantly lowered, benefiting not only vegetable and sugarcane growers but the expanding suburbs of Miami, a settlement that became a city in the generation following the arrival there in 1896 of the Florida East Coast Railroad. The transport revolution whose trains had brought the lumber industry to inaccessible areas a generation before, now—aided by a new device, the automobile—helped to bring northerners and midwesterners to central and lower Florida, to its groves, drained swampland, and cities. Against many difficulties, commercial farming came to the Everglades about 1915, and developed slowly during the 1920s. In 1926 and again in 1928 hurricanes pushed storm surges over protection levees on the south shore of Lake Okeechobee, killing some 2,500 people in all. Despite the storms and a drastic deflation in the real estate market, however, the Dade County phenomenon endured, strange to the South then, but soon to grow more familiar.[30]

Of at least as great significance was the discovery of oil in Texas in 1901. Other oil discoveries in Arkansas, Oklahoma, and Louisiana followed.

The growth of the petrochemical industry followed the rise of the automobile, and the rise of Houston that of the industry. In 1920 Houston was a lively town of 138,000; in 1930 it was nearing 300,000 and was launched on its career as a smaller regional Chicago. Entertaining parallels could be drawn between the two cities of the southern borderlands, Miami and Houston. Even in Indian times, south Florida and Texas had not shared in the Mississippian culture, nor had they been fully a part of the Old South system. Houston belonged as much to the West as to the South. As they prospered, both cities became the goal of immigrants from the North. While Miami grew around citrus, tourism, and light industry, Houston focused on oil and its satellite concerns, on agriculture, and on shipping.

As the South would soon discover, the relationship of industry and conservation is complex. It has been argued that industrial technology, even more than other forms of human effort, depends upon the natural environment. Its very impact makes for dependence, as companies find out when they overload natural waste-removal systems. The nature and location of resources, including climate and landform, shape the possibilities of production at every step. The petrochemical industry in particular rests upon layered environments, ancient and modern; considered as an alimentary system, it devours an ancient environment and discharges its wastes upon the present one. The transformation of Texas's Buffalo Bayou into the Houston Ship Channel, a remarkably poisonous body of water, might serve as a convenient symbol of the new problems which urbanization and the oil Golconda entailed.[31] But amid the pervasive regional poverty, oil shone for many as a new Gold Rush, and many a thriving southwesterner, like the young Louisianian Huey P. Long, invested in oil hoping that "I might one day be mentioned among the millionaires."[32]

Progress and poverty were the poles of the regional mind. Following an example given by New Jersey, several southern states set up agencies similar to chambers of commerce to promote immigration and industry; these bodies took their coloration from the Progressive conservation movement and talked of conservation and development even-handedly, in the best modern way. North Carolina's Department of Conservation and Development (established in 1925) reported that it was "in the nature of a State chamber of commerce" whose duty was to "point out in broad terms existing conditions for the guidance of trade bodies in promoting the growth of their communities and of the State at large." Virginia set up the same sort of agency in 1926, and soon development commissions—the Alabama Industrial Development Board, the South Carolina Natural Resources Commission—undertook to bring whatever story of untapped resources and beckoning opportunity they could uncover to the attention of potential customers. The work was made a constitutional function in Alabama, Louisiana, North Carolina, Texas, and Virginia.[33]

But this is only to say that conservation in the South, as elsewhere, often meant simply long-sighted and cost-effective development. In the 1920s, preventing the loss of natural gas, often burned at the wellhead, drew the attention of President Calvin Coolidge's Oil Conservation Board. State legislation against waste followed; Huey Long made providing natural gas to homeowners a feature of his early campaigns. The arrival of the Depression only intensified the need to exert control, this time in the interest of reducing production. In 1931 a compact of the producing states was "temporarily enforced by martial law in Texas and Oklahoma."[34] This was conservation at the gut level of commercial need or desperation. Throughout the 1920s, coordinating efforts at the national and state levels joined science to boosterism, conservation to production, in forms that adumbrated the conscription of resources during the Depression and the war years that lay ahead.

Conservation as the partner of development was nowhere more boldly pursued than in the nation's river basins. And nowhere were greater sums expended, greater risks incurred, and more initiatives taken than in the alluvial valley of the Mississippi.

The twentieth century found the Mississippi River Commission reacting to rising flood heights by repeatedly increasing the elevation and cross-section of the mainline levees. At the same time, the Commission was extending and unifying its control up the river and into the tributary basins. Levee building continued its technical development, as levee machines replaced the primitive wheelbarrow and were supplanted in turn by motorized earth-moving equipment. But the narrower sort of technical progress meant little in this case; what was needed was a new approach to the problems of the river. Citizens' groups in the lower valley, reacting to the danger which the rising floodline posed to immovable property, such as the New Orleans docks, began to press for new controls by spillways or floodways to supplement the levee system. Until 1927, however, the Commission—backed by many powerful interest groups and floodplain solons like Louisiana's Senator Joseph E. Ransdell—kept to the levees-only dogma.

Meanwhile the federal government was responding to local pressures by amplifying its commitment to flood control. The Flood Control Act of 1917, committing the government to such work on the Mississippi and Sacramento rivers, was sometimes referred to as the first flood control act. So it was, in theory. The law by no means represented the first federal work in the field, but it did break new ground in frankness; it did not begin the era of federal flood control, but it ended the era of federal subterfuge. The Rivers and Harbors Act of 1925, on the other hand, had genuinely new features. Laying aside bureaucratic struggles over turf that had brought the Corps of Engineers into earlier frays against comprehensive river basin development, the 1925 act provided that the Corps itself should plan such development for many of the nation's rivers. The Corps responded handsomely, providing blueprints for work that was done later during the Depression, when the need to relieve unemployment brought the Corps and the relief agencies together in many a major endeavor.

Congress also resolved to take a new look at the Mississippi. Plans for various reforms were under study by the MRC and several Corps of Engineers boards were studying alternative methods of control when, in 1927, a major disaster hit the valley. Preceded by very high guage readings in the autumn of 1926, synchronized spring rises in the major tributary basins combined with heavy rains in the valley itself to create one of the great floods of history. By mid-April, lowlands were flooding, thousands of people were homeless, and a dozen or so were dead. The Arkansas River, raised by cloudbursts in Kansas, flooded some 15,000 acres. Refugees poured into towns and hastily established camps on high land, where contagious diseases promptly broke out. For the first time, a mainline levee built to Commission standards failed at Mound Landing, Mississippi, flooding several million acres. In the three most southerly states there were more than forty crevasses. The federal government organized in unprecedented fashion for relief, as Secretary of Commerce Herbert Hoover took charge of a presidential commission; the Chief of Engineers went to Memphis to direct the floodfight, and seven federal agencies worked in an integrated relief effort. After massive destruction that made some 700,000 people refugees, the flood began to recede; below New Orleans, the state of Louisiana, on Commission advice, cut a levee to relieve pressure on those about the city. In this manner the final crest passed into Lake Borgne and the Gulf of Mexico.[35]

The ride on the tiger had been a rough one. There was not, of course, any question of the valley's inhabitants trekking out of the floodplain, or putting their houses (and cities) on stilts, or learning to live at subsistence level. Nor was there any question of simply raising the levees yet again. "The cost of levees on the Mississippi," wrote a later Chief of Engineers, "increases more rapidly than the square of their height, and the destructiveness of a crevasse increases almost in like proportion. . . . All conditions demand levees of limited height, and the limit is soon reached."[36]

Less easy to admit was the fact that the levee system, then a bit over 200 years old, had played a major role in the calamity just past. The levees had not, of course, caused the flood, but they had caused much of the death and social upheaval that resulted from it by making heavy settlement in the floodplain possible and by creating a false sense of confidence among the people of the valley. In fact, they had changed the whole character of floods: because levees stopped normal rises from topping the banks, they made floods rarer; because the leveed channel was smoother, the river ran faster, and flood crests were hurried along to the sea. The levees also insured that when floods did occur they would be more catastrophic and that the river's prime work of land-building would cease. In a sense, by building the levees, the human societies along the riverbanks had invented a new kind of river—one which, in turn, required control. Still more distressing was the fact that the era of major improvement just past had worsened the system's defects by improving its technology. In the forty-eight years since the founding of the Commission, a great deal of money, brains, and work had gone into perfecting the levee system, with results that were, at least for the moment, painful to contemplate.

The answers of the time were more money, reorganization, increased knowledge. Congress funded the Mississippi River and Tributaries Project, reduced the Commission to an appendage of the Office of the Chief of Engineers, authorized the building of floodways and a spillway, and set up a major hydraulic laboratory, the Waterways Experiment Station, at Vicksburg. The basic technical feature of the new approach was to provide controlled-access channels parallel to the river for use in great floods only, plus an artificial outlet to Lake Pontchartrain above New Orleans to protect the city. In effect, the plan sought to recreate one major effect of the floodplain which the levees had cut off, by allowing the river in time of flood to expand sideways as well as up. This in turn compelled the development of the Atachafalaya River in Louisiana, a major natural distributary of the Mississippi, as a floodway system, with an array of environmental consequences to be considered later. The working out of the Jadwin Plan, however, belonged to the 1930s, another in the considerable list of major projects for which the 1920s had provided the drawing-board.

By the end of Hoover's administration, scientific research within the government had progressed impressively. The agricultural scientist Hugh Hammond Bennett had accumulated the information on which he would base his Soil Conservation Service. The Bureau of Fisheries had greatly expanded knowledge of finny habits and diseases. The Bureau of Mines had devised new methods of extraction and refining; the Forest Service had completed a nationwide timber survey. The bases of New Deal resource policies had in many cases been laid under the engineer president, a service for which Hoover has received only belated recognition.[37]

While by no means a seamless web, national resource policy had been woven with a fairly steady hand during the early decades of the century in response to imperatives which neither political party was able to ignore. While creating great opportunities for the sportsman and the urbanite with atavistic impulses, the conservation movement had provided occupation for scientists and bureaucrats, profits for businessmen, and an elevator for the politically ambitious. It had, in short, succeeded in the way that such movements do succeed, by meaning many things to many different people and by providing some advantages to almost all. For the South, its major contributions were still to come as resource policy became an aspect of Depression-era politics. In the coming decades the national government, working usually through and with local authorities, and often guided by southern lawmakers into appropriate statutory channels, would seek to transform the region's natural and human resources by a multitude of programs.

8

The Transformation Begins

IF THE DEPRESSION DECADE HAD BEEN MERELY A TIME OF INCREASED PRIVATION, it would rate little attention in any history of the South, including this one. Instead, for the section perhaps even more than for the nation, the decade proved to be the wellhead of remarkable changes in government and society. Of these changes, many were reinforced and some perverted by the war that followed.

Despite the fact that the Roosevelt administration lavished far more rhetoric than money upon the South's problems, its varied and paradoxical programs helped to launch a many-sided revolution. Some New Deal policies represented a holding action, sustaining life in a period of want. But others helped southerners to change old and bad habits in the treatment of their land. Improved delivery systems were provided for the discoveries of maturing medical and agricultural sciences. Most basic, perhaps, were the programs that helped to change the nature of southern farming, at grave cost to the poorest workers on the land. When wartime demands for soldiers, workers, energy, and raw materials followed, some basic and seemingly immutable facts of regional life began to change beyond recognition.

The long process of transforming the lower Mississippi River from a half-wild to a half-disciplined stream made great advances during the 1930s, but also encountered great problems.[1] The largest of the southern floodways was to utilize the Atchafalaya River, a natural distributary of the Mississippi which flows through a marshy, often primeval region of central Louisiana to the Gulf of Mexico. The Corps of Engineers entered in haste upon the project and soon found itself entangled in problems of politics and land use, engineering and ecology that a generation was not to solve. The primeval look of the swamp was not deceiving; this was new land, much of it formed during historic times, as the Atchafalaya itself had been. To impose human works where nature's were as yet incomplete proved no easy task. For this area, the Corps's plans called for creating an efficient channel and, by means of guide levees, three parallel floodways. Until the end of the 1960s little serious attention was paid to the effects all this would have upon wildlife.[2]

Of great moment to the human residents of lower Louisiana was the possible diversion of the Mississippi through the new channel. Six times in geologically recent ages the river had shifted its outlet to the sea as aggradation decreased the slope of its established course. Since the time of the Louisiana Purchase, many observers had foreseen that the Atchafalaya, as a natural outlet of the Mississippi, might one day become the main channel of the river. Long before the 1927 flood and the Mississippi River and Tributaries project, Louisianians pursuing their own ends, mostly commercial, had done all they could to increase the flow of the stream. The Corps therefore acted in line with well-established local tradition when it destroyed existing sill dams and embarked on its own dredging program. By 1940 the Atchafalaya provided the great river with a three-to-one advantage in slope over the existing channel past New Orleans.[3]

Meanwhile the swamp proved as fecund of difficulties as it was of life. The basin was a kind of environmental Looking-Glass Country where the engineers must run as hard as they could in order to stand still. Heavy silt aggraded the channel, raising flood levels, while the guide levees built to define the limits of the flood sank on foundations of mucky, unconsolidated soil. But political realities demanded that the effort continue against all obstacles. Good luck intervened in at least one form for the hardworking men on the engineer quarter-boats. The floodways were not needed for forty years, and meantime the surgical resection of central Louisiana had the advantage of considerable sums of money and constantly improving techniques.

On the main channel of the Mississippi, cautious innovations brought quick though limited improvements. Building a spillway above New Orleans provided the city protection by inserting into the levee line a weir whose floodgates—actually balks of wood called needles—could be pulled one by one to provide a controlled outflow between guide levees to Lake Pontchartrain. For some centuries or millennia the river had been trying to force itself an outlet at the bend named Bonnet Carré; engineering did the job briskly and provided a spigot under human control. In the later 1930s Brigadier General Harley B. Ferguson, the Corps's division engineer at Vicksburg, began experimenting with artificial cutoffs, that is, dredging across meander loops to shorten the river channel and improve its efficiency. A violation of standing engineering dogma, cutoffs proved to work when done cautiously in response to local conditions. Though temporary in their nature (the tendency of the river to meander was not to be reversed) cutoffs meant that the main channel was better able to handle floods, and in consequence that levees need not be so high or diversion works so extensive. Such craftsmanship had much to offer the communities of the floodplain.[4]

While work on the Mississippi floodplan went forward, federal flood control was extended beyond the limits established in the act of 1917. The Rivers and Harbors Act of 1936 launched a nationwide experiment in environmental transformation of incalculable consequences. Comprehensive development was now a national goal, and the Depression provided an army of the unemployed to labor on the projects which resulted. Over the work, like an art deco eagle, hovered the classic liberal principle of the greatest good for the greatest number—now embodied in the cost-benefit ratio—as guide, rationale, and justification for the whole.

The consequences for the flowing waters of America, for the land they divided, and for the human and nonhuman communities that used them, were immense and full of paradox. The restraint of floods, the production of energy, and the improvement of transport would result as planned. The flooding of farm and forest land artificially through impoundments would alter much of the landscape and, for better or worse, the lifestyle of millions. A rising population in floodplains would, in turn, increase flood losses in dollar terms.[5] Elaborated from unspoken assumptions that were largely commercial and technocratic, river basin development impacted in the most varied ways, enabling depressed regions to live anew even as its environmental consequences demonstrated that an efficient and profitable river was not always a living river.

On the Mississippi there were some immediate benefits from the new control projects. In January 1937 one of the greatest of recorded floods started down the Mississippi. A flow of nearly one and a half million cubic feet per second was recorded at Cairo, Illinois. In the alluvial valley, some gauge readings exceeded those of 1927. One of the new floodways, in Missouri, was opened by dynamite. Backwater areas flooded, the Red Cross and National Guard were called out, and the Public Works Administration (PWA) and the Civilian Conservation Corps (CCC) provided labor for sentry and maintenance work along the levees. On February 18 Bonnet Carré spillway was opened, diverting over 200,000 cubic feet per second from the swollen channel. By March 16 all floodgates were closed again. The spillway had worked, the system as a whole demonstrating the potential of engineering when it assisted or mimicked nature.[6]

Much suffering had come with the high water, yet comparison with the events of 1927—made instinctively rather than accurately, considering the major differences between the floods in scope, form, and duration— pointed in one direction. With all its faults the flood control plan in its first decade had gone a long way toward making the alluvial valley a safer place to inhabit. Though the engineers who were in closest contact with the Mississippi were usually least apt to speak of permanence in their work, the lesson of their achievement was drawn more simply by their bureaucratic superiors, by congressmen, and by the public. Man had disciplined one of the world's great rivers. Perhaps he could do with the others as he chose.

It was another southern river, however, that provided the most remarkable demonstration of comprehensive basin development under the Tennessee Valley Authority. The valley itself is a great arc whose main channel and tributaries drain much of Tennessee, western Virginia and North Carolina, northern Georgia and Alabama, and western Kentucky. Born in a mood of fervid experimentation, the Authority evolved within a decade to a solid and conservative part of the regional establishment. Along the way it became an educator in agriculture, public health, and resource management, a contributor to the increasing prosperity of its valley and region, and a focus of cooperative effort between state, local, and federal agencies.[7]

Muscle Shoals was an obstruction to river commerce that had long been a target of local boosters. Its twentieth-century development dated to the period of World War I, when concern over the possibility of Chilean nitrates being cut off—nitrates were essential for manufacturing both explosives and fertilizers—led the federal government to authorize the construction in northwestern Alabama of one (later two) nitrate plants, one at Muscle Shoals, and of the Wilson Dam as a prospective power source. The idea was to provide explosives in war and nitrate fertilizers, much favored by southern farmers as a chemical fix for depleted soils, in time of peace. During the 1920s, however, Muscle Shoals became the bone of contention between advocates of private power development, backing an offer from Henry Ford to relieve the government of its investment there, and those of public power, led by George Norris. Norris's aim, however, went beyond power alone to embrace a long-discussed goal of conservationists—the comprehensive, integrated development of a river.

Franklin D. Roosevelt took up the issue in the election of 1932. In 1933–1934, under the first TVA Act, a fine frenzy of experimental beginnings was made in the teeth of tenacious opposition by private power companies alarmed at federal competition. An array of federal agencies moved into assist at the parturition of TVA. The Bureau of Reclamation designed the first dams, sale of whose power was to be the moneymaking aspect of the Authority. The Corps of Engineers advised on flood control and navigation. Under the concept of multipurpose development, the many aspects of a river valley were to be treated interdependently, a revised ecology, so to speak, bent to human purposes. Erosion must be attacked to control siltation and extend the life of the reservoirs; this implied agricultural education for the farmers and the reforestation of watersheds. The transformation of a running river into the impoundments necessary for navigation and to create the artificial waterfalls that ran the power turbines, implied extensive environmental change which also had to be managed. There were obvious dangers that still water would encourage the breeding of mosquitoes in a region where malaria was already endemic. TVA therefore implied public health work. The impact of dams on fishlife and indeed on the whole ecology of the natural river was little known, but could only be considerable. Systematic studies by the TVA revised basic ideas of reservoir biology and made possible year-round game fishing; thus recreation became part of development. Here was the gardening of nature on a grand scale, demanding the ability to think across the conventional boundaries of the sciences and to act across those which divided the federal agencies, with their carefully delineated turfs.

In practice, however, the fundamental orientation of TVA was and remained economic. This character was soon intensified by court challenges and internal struggles which, by the late 1930s, led to the triumph of practical administrators whose aim was to fulfill economic goals and win the support of local interests. Keyed to development, the TVA of David Lilienthal and Harcourt Morgan came to mean in essence navigability and cheap power. Other functions were transferred to traditional federal or state agencies, where necessary, to decrease opposition. For example, the Forestry Relations Department ceased to acquire land for reforestation and transferred its surplus land to state conservation agencies, the U.S. Forest Service, the National Park Service, or the Fish and Wildlife Service.

Yet much of the character of a unique regional authority survived. The Forestry Relations Department's own functions remained broad. Of its three divisions, Watershed Protection concerned itself with reforestation, erosion control, and fire protection. Forestry Investigations was the research arm. Biological Readjustment was preoccupied with fish and wildlife. Under the Department too were thirty camps of the Civilian Conservation Corps, the New Deal's experiment in healthy work and disciplined living for idle youths. With these forces, TVA pursued an impressive program of conservation. Two forest nurseries were set up and young trees made available for private land at no charge—the CCC did the work of planting. Landowners had only to ask for help and agree to protect the plantings from damage by fire and grazing. From an apathetic first reception, interest grew quickly, and between 1934 and 1936, 18 million trees were set out on 11,560 acres located in some 1,700 private properties. Recognizing that the CCC would not last forever, the Authority involved landowners too in the planting program.[8]

Conservation programs faced challenges from within TVA itself. Chief

Forester Willis M. Baker found his work cut out for him when he attempted to convince Authority engineers who operated the dams and reservoirs to avoid the sudden fluctuations in water levels that destroyed aquatic life. Such difficulties with engineers might have been anticipated; however, Baker also faced some sharp disputes with the protectors of human life. The Health and Safety Department, pursuing its public health mission, sprayed oil and poisons along the reservoirs to kill *Anopheles;* naturally they killed fish and wildlife, as well.[9]

The public health people had their own achievements to record. The Authority took the position that while it had no obligation to eliminate malaria in the valley, it was obligated not to increase its incidence. Other diseases also followed TVA construction crews and had to be dealt with. The usual ills of new and raw towns, their outbreaks tediously repeated since the time of Jamestown, appeared again, as typhoid, highly endemic in the valley, increased at the construction sites. The Authority undertook to render specific services under contract to county health officers, mainly in malaria control, sanitation, and epidemiology. It provided health services also for its own employees, and its Tennessee program ultimately extended to communicable diseases, establishment of tuberculosis clinics, and a variety of community health programs.[10]

TVA was not, however, an eleemosynary institution. It was a product of the Progressive tradition of resource management, and more specifically of the Depression decade. The business of America was economic survival, and TVA was interested basically in the development of the river, in power, and in agriculture. Its basic achievement was reflected in the fact that, following TVA's creation, economic growth in the valley generally outpaced that of the South and the nation. Commerce on the river rose, industry migrated to the valley, and the increased availability of energy showed its usual association with rising levels of prosperity. Over the thirty years following the creation of TVA, per capita income rose from about one-fifth to about two-thirds the national average. TVA had not been a fairy godmother, but evidence is persuasive that it was an essential catalyst in much of this change.

Fundamental was the transformation of valley agriculture, to which the Authority contributed in a variety of ways. A region of poorish soils lacking many essential chemicals (calcium and phosphorus primarily, with less marked deficiencies of potassium, nitrogen, and organic matter), and extensively eroded, the valley told an all-too-familiar southern story of a mediocre natural dower worsened by long abuse. Rural population density was high, as elsewhere in the Appalachian uplands, though towns were few. Cotton farming dominated the landscape, and there was little effort to adjust farming to the soils and environment. Warmth and heavy rainfall were the prime natural advantages of the region, as well as being burdens which encouraged the vectors of disease and produced erosion.

At first the TVA offered local farmers little except the nitrate fertilizers from Muscle Shoals, where the whole development had begun. But experts perceived the limitations of the old southern dependence on fertilizers to improve poor soils: increased row-crop agriculture, more erosion, and consequent problems for the Authority's impoundments and navigation works. The TVA began to encourage crop rotation and use of Tennessee's natural phosphate rock, processed at Muscle Shoals, to correct the phosphorus deficiency and make the growth of legumes possible. Demonstration farms became the favored method of education, and the Authority also worked through agricultural extension services and county agents. The general shift in valley agriculture over the generation that followed—the decrease in cotton and corn, the increased production of mixed crops and dairy cattle—was not caused by TVA alone, for the whole South was moving in the same direction. It did, however, certify to the wisdom of the course the Authority's agricultural experts had marked out.[11]

By the opening of World War II, TVA was a solid success. Its evolution inspired varied emotions: David Lilienthal saw it as democracy on the march, while utopians like its onetime chief Arthur Morgan and the social experimenter Rexford Tugwell were less complimentary. The latter declared that "TVA is more an example of democracy in retreat that democracy on the march." Critics on the right continued to deplore an active, interventionist government, particularly one which produced cheap power. Critics on the left as vigorously declared that "from 1936 on the TVA should have been called the Tennessee Valley Power Production and Flood Control Corporation," and scored Lilienthal's alleged overreaction to court challenges from private power companies, and Harcourt Morgan's supposed subservience to local interests and the American Farm Bureau. Through TVA a good deal of money and a variety of expert knowledge—in engineering and commerce, agriculture and public health, forestry and conservation—had been applied to ends that comported with a capitalist society and the aims of the local elites of the Tennessee valley.[12]

The nation at war found uses for TVA capabilities which were still more brutally pragmatic than the peacetime enhancement of private development and profits. Forestry took on a wartime cast. TVA foresters worked with industries and landowners to spread the Pinchotian gospel of sustained production and multiple-use management; one result of their work was increased production of forest products for the war effort. There were direct military applications: TVA's electric-furnace process for producing fertilizers was also the most efficient for turning out munitions-grade phosphorus. But the most remarkable contribution by TVA to the struggle was the use of its power for the Manhattan project at Oak Ridge. Military purposes had been involved in the Muscle Shoals development from 1916, but this rather transcended what anyone had expected.[13]

Yet intelligence had been applied through TVA with marked and fair success to resource conservation as well as development. Nation, state, and locality had cooperated to the substantial betterment of all. The early history of the Authority was full of irony, but what large endeavor can be cited whose history is other than a record of unexpected denouements? The South in the Depression years could probably have used a few more TVAs. One was interesting, but not enough.

Lacking multiple TVAs, the rest of the South got along as best it could under general programs. The region was something of a stepchild in terms of the total amount of money it received during the 1930s, but this did not prevent the New Deal programs from being a bonanza in an area of such widespread poverty. Certain favored groups within the South got very substantial dollar totals, to say nothing of loans of labor, credit, and expertise. The very poor benefited also; at any rate, they were helped to survive.

Basic to public health work were efforts to improve the diet of the rural poor. Since early in the century, experimental studies of nutrition by the Public Health Service and the Department of Agriculture had proliferated, including practical investigations into the vitamin content of cheap and easily accessible green foodstuffs such as collard, mustard, and kale. The well-developed network of landgrant colleges, experiment stations, and county agents joined public health workers in urging on the rural population the addition of increased quantities of greens to the prevailing southern diet, with its overdependence on pork and corn. The extension service spread the discovery of Dr. Joseph Goldberger of the Public Health Service that an enriched diet could prevent and cure pellagra. During the Depression, direct food aid to hungry southerners helped to stave off want for many. Food was important, not only in combating the nutritional diseases, but the building strength to resist infectious ills, as well. Yet there can be little doubt that New Deal efforts were stopgaps only, or that the southern poor, especially the less accessible rural poor, suffered through the 1930s from continued and probably intensified ill-health. In 1938 medical examinations of 565 black and white sharecroppers in Laurens and Oglethorpe counties, Georgia, showed that 121 children had rickets, more than 40 people were anemic, and 14 had pellagra. Tuberculosis also was common.

Pellagra was unique. Identified in Europe during the eighteenth century, the disease was not certainly present in the United States until the early twentieth century, though there is some evidence of seasonal outbreaks among slaves in the Old South, especially during the late winter and early spring when nutrition was most apt to be insufficient. ("Pellagra," said Goldberger, "is a sort of spring crop following a lean winter diet.") The illness was a dissembler, manifesting itself primarily in three seemingly unrelated symptoms: dermatitis, which was often mistaken for sunburn; diarrhea, which was common from a variety of causes; and insanity. Well into the twentieth century physicians were uncertain whether pellagra was a deficiency disease or one caused by moldy corn, or whether it was infectious and spread by an insect vector. Whether in northern Italy, in eastern Europe, in Egypt, or in southern Africa, it was associated with areas where the poor depended heavily on maize as the mainstay of their diet. By the early years of the century, public health investigators, led by the South Carolina Board of Health, had identified it as a widespread southern affliction. Its spread was symptomatic of the increasingly impoverished diet brought on by the obsessive culture of cotton, the lack of home-grown vegetables and milk, and the depletion of wild animals as a source of fresh meat. A number of deficiency conditions resulted, of which pellagra was the most famous. The disease apparently declined during 1929–1932, when the cotton market hit bottom and many farmers turned to food growing. Pellagra rose again in the middle 1930s as sharecroppers were thrust off the land, but its demise was now at hand. In 1937 researchers identified nicotinic acid as the critical deficiency, and the deliberate enrichment of bread began in the early 1940s. Even more to the point, wartime jobs in industry, the daily three squares for those in the armed services, and all the other changes of the time spelled drastically improved nutrition for millions.

Pellagra's death rate tumbled in a few years. In 1940 over 2,000 Americans died from the disease; in 1945 there were 865 victims. The case rate was halved during the same years. Even those devoted to the traditional rural diet found, in the words of a medical historian, that it had become "difficult to avoid vitamin B." By the most artificial means, an industrial society achieved a form of dietary balance which the years of monoculture had lost.[14]

As for malaria, that durable illness continued its course during the 1930s, the last full decade during which it was to affect much of the South, but under increasing pressure from a host of enemies. Relief agencies turned out manpower for a variety of health programs; in the South, some 33,655 miles of drainage ditches were dug, eliminating an estimated 544,414 acres of breeding grounds for *Anopheles*. The practice of strengthening local agencies rather than superimposing new ones brought good results. Malaria survey and control personnel were added to state health departments, and federal funds provided to increase the number of local health departments. Works Progress Administration assistance continued until late 1941, after which the system of survey and control teams, together with many of those who had worked in them, was taken over by the armed forces to aid campaigns in the malarial regions of the Pacific, North Africa, and the Mediterranean.[15]

In addition, the 1930s saw chemists add two decisive weapons to the armory against malaria. Quinacrine, or atabrine, a synthetic drug more powerful than quinine, was tested in Georgia in 1932 and by World War II was in mass production for the armed forces. In 1939 the Swiss experimenter Paul Muller demonstrated the insecticidal properties of an organic chemical, dichloro-diphenyl-trichloroethane. By 1942 both the army and the Rockefeller Foundation were testing its properties in the United States. Under the acronym DDT, the insecticide was to have an extraordinary effect on the disease environment, followed by that of agriculture and of wildlife.

Yet for all that was done and discovered, the malaria battle during the 1930s, like the anti-hookworm fight, yielded no decisive victory. In 1935, 89 percent of the nation's 4,435 malaria deaths were recorded in the South; the lower South as usual had more than the upper, rural areas more than towns. Malaria remained a part of the rural lifestyle in much of the South. The disease survived for a multitude of oppressively familiar reasons: because the climate was warm and the area large, because the populace was heavily seeded with parasites, and because many could not afford screens for their houses. A rhythmic surge in incidence occurred in the middle 1930s, its peak lifted by Depression-era hunger and poverty. Tenants on a Louisiana plantation pressed their overseer for aid, not only to get food and clothing, but for protection against fever: "The no the [they know they] are going to be sick," reported the overseer, "because the m sceaters [mosquitoes] are eating them up and [they] ant able to get A [mosquito] bar [netting]."[16]

In 1940 the army again began to move in. Many camps for training the new draftees were located in highly endemic areas. As the nation shifted to a war footing, munitions plants, airplane factories, and shipyards began to locate in a region of ample space and cheap, available manpower. Laborers vanished from WPA rosters; malaria-control projects had to be abandoned. Meanwhile civilians and soldiers utterly unfamiliar with the disease crowded together in and around the camps and factories, many in close contact with infected southerners. The possibilities for a worse 1898 were evident.

Hence the army—first the Quartermaster Corps, later the Corps of Engineers—took charge of mosquito control on the posts. Offpost the Public Health Service combined forces with state and local authorities in a campaign coordinated by its Surgeon General. In February 1942 an Office of Malaria Control in War Areas (MCWA) was established with headquarters at Atlanta. The antimalaria team—a physician, and entomologist, and an engineer—formed the basic unit in the fight, inspecting its area in borrowed army vehicles. Vector control by drainage and spraying was the keynote of the campaign. Nationwide the MCWA spent about $31 million on control, the military services another $11.5 million on cantonment projects. The lack of epidemic outbreaks like those of the Spanish-American War marked the success of their efforts. Malaria could be suppressed in small areas by careful policing and firm discipline. But its career as an endemic disease of southerners had not yet run its course.[17]

New Deal policies also helped to reshape the conservation of forests and wildlife. Much credit deservedly went to the CCC, but much also belonged to the government's policy of acquiring cutover and badly eroded land and turning it into parks or national forests. A system of state parks also grew up, as state governments followed the national model in order to exploit CCC funds and labor. Wildlife refuges expanded much more slowly, but—aided by the ubiquitous tree-planters—some progress was recorded here also.

At issue fundamentally was the protection and restoration of the land. Under relentless abuse by logging, burning, and grazing, forest land had passed along with much former cropland into the submarginal category. Critical was the inability of such land to bear vegetation to the extent necessary to prevent further destruction. Soil and wildlife conservation, as Hugh Hammond Bennett pointed out, linked in the necessity to restore groundcover.[18] That was where the CCC came in.

Combining conservation with unemployment relief, the agency more than any other was FDR's baby. Though most of the camps were in the West (the region which actually received the greatest share of the New Deal largesse), from its first camp, set up at Luray, Virginia, in April 1933, the CCC became part of the southern scene, as well. A welcome part, all in all, for the camps brought money as well as workers into the places where they were established, and the hard labor and back-to-nature aspects of the experiment pleased even conservatives. Though under military officers, the camps worked for various civil agencies, especially the Departments of Agriculture and Interior, and also for the state park and conservation agencies. Much work was done on private land. Forest work included fire control, building of roads and some dams, and pest control. The most important was reforestation. Not only did CCC workers do extensive planting—some 570 million trees in the national forests alone by mid-1936—, they encouraged growth by thinning overcrowded timber stands.[19]

Second to the Forest Service among Agriculture Department bureaus using the CCC was the Soil Conservation Service. Most SCS work was done in the South and West, where workers checked and healed gullies, terraced sloping land, and contour-planted trees to prevent erosion.

Under the Bureau of the Biological Survey, the CCC developed submarginal land as refuges and stocked streams with fish they had reared. The "Arkansas floating camp" lived on houseboats while working on waterfowl refuges in marshes, rivers, and bayous. For the Interior Department the CCC worked on national parks, mainly to provide access and recreation for park visitors; the department also purchased land to make new parks, the largest acquisition being Big Bend National Park in Texas.

Among the useful (and profitable) contributions of the CCC was the planting of 168,721 acres of worn-out farmland to trees. The professional woodsmen hired to guide the planting taught that trees could be a good crop on land that row crops would ruin, but that they must be planted intelligently, the right species for the right combination of site, climate, and soil. Follow-up studies of pine plantations in Alabama, Mississippi, North Carolina, Tennessee, and Virginia in 1953, twenty years after planting, indicated that landowners were receiving returns of 12 percent annually from thinning alone on their modest first investments. Trees could not be grown on all land, but to those who grew them well they could bring some very practical rewards.[20]

Such achievements led some foresters to estimate that eight years of CCC work advanced state and national conservation by twenty-five to forty years. Rates of planting were highest in the South, about 1,000 per man per day. The CCC also brought the southern forests a measure of protection against uncontrolled burning. Overall, the agency's success was measured at least in part by the conversion of its opponents. In Louisiana the redoubtable Senator Huey P. Long, as part of his guerrilla warfare against the New Deal, had denounced the original CCC legislation as "the sapling bill." At first work proceeded with little cooperation from the state, which in any case was without a park system of its own. Four camps aided state work on levees, while most of the remaining twenty did forestry work on federal or private land. Kisatchie National Forest received two. After Long's death in 1935 the atmosphere warmed considerably, as his successors moved to mend their fences with the federal administration. After passage of the federal Park, Parkway and Recreation Act (1936), Louisiana officials worked with Harold Ickes's Interior Department to develop a system of state parks, including Chemin-a-Haut, Tchefuncte, and Chicot (1936, 1937, and 1938, respectively). By 1937 Louisiana also "led the South and was second in the nation in the amount of CCC reforestation work on federal forest lands," with the region's largest tree nursery in the Kisatchie Forest near Alexandria.[21]

It should be noted that government efforts, though large, were never the whole story. The evolution of the forest products industry had a conservationist aspect. A breakthrough had come with the spread of pulpmills and papermills which provided a market for trees of small size. Though the industry dated from the late nineteenth century, a major development for the South was the building of a large sulfate-process mill at Savannah in 1936. Within five years another thirty mills were constructed. With their expensive machinery, such mills, unlike the fly-by-night sawmills, were permanent establishments requiring permanent sources of supply. In the winter of 1936–1937 industry leaders and foresters drew up rules for cutting with the aim of preserving and improving forest stock. Trees below a certain size were to be preserved, seed trees left after cutting, and logging confined as far as possible to damaged trees or those in statnds which required thinning. Enforcement of such rules was another matter. The Southern Pulpwood Conservation Association was formed in 1939, but not all the companies joined; the association pursued the goal of insuring a durable supply by propaganda and by distributing seedlings for reforestation, apparently with considerable effect on corporations but little on private owners interested only in a quick return.[22]

In 1940 the industry spread to Texas, as the manufacturing of newsprint began at the Southland mill in Lufkin. In Bogalusa, Louisiana, several phases in the history of the industry met. The town grew up around lumbermills which attacked the local yellow pine forest in 1908. In the early twentieth century, practices were as wasteful as any to be found in the South. In 1916 the first pulp and papermills moved in, aiming to use the wasted forest byproducts. Soon the fact became starkly apparent that the industry had no future without a secure source of raw material. A program of reseeding and hand planting was launched, using "worn out Model T Ford axles and broom handles" as dibbles to make holes for the seedlings. In 1937 the lumbermill closed for lack of trees, but the pulp and papermills continued replanting and expanded their operations. This example suggests why the South had so much ruined forestland for the CCC to work with, and also why manufacturers of many sorts learned the lessons of conservation in the school of hard knocks. It was also why many southerners saw in the course of a generation—as in the South Carolina up-country described by T.S. Buie in an article entitled "From Pines to Pines"—the land they knew pass from mixed forest and farmland, through a period of devastation, and back via government action or industrial conservation to forestland.[23]

The replanting of forests also meant the spread of edgelands between woods and fields where wildlife found cover and food. Attacks on erosion meant some improvement in the control of silt, and hence increased chances of survival for fish, which silt destroyed by burying eggs and occluding light. More direct in their influence on wildlife were the refuges established during the same period. State conservation programs were essentially sportsman-oriented, featuring the usual Progressive pattern of regulated hunting, plus predator control designed to please stockmen and farmers. State and federal refuges, on the other hand, featured positive efforts not only to protect wild species but to increase their numbers. State refuges had been set up in Louisiana in 1905 and Alabama in 1907. In 1929 Congress had authorized the purchase of land for migratory bird refuges, and in May 1933 President Roosevelt approved the first CCC camps for wildlife refuges in Maryland, North Carolina, and Florida. The next year the Migratory Bird Hunting Stamp Act provided funds for purchasing and developing land. With workers available, the government found money in several pockets—"duck stamp" funds, submarginal land bought up by the Resettlement Administration, and others—to set up a number of southern refuges. Among these were the Carolina Sandhills National Wildlife Refuge, where deliberate restocking of the nearly extinct Carolina beaver began; the Piedmont refuge near Macon, Georgia, in a region of badly eroded hill country; the Noxabee refuge on similar land near Starkville, Mississippi; the extensive Kentucky Woodlands refuge; and the Okefenokee, set up during 1937 in part to protect migrating waterfowl and big game.[24]

The Peaceable Kingdom was not, however, at hand. Conflicts regularly developed between wildlife managers and foresters, each specialty having its own view on the question of whether woodland should be treated as a timber resource or as a habitat. There were also conflicts between state and federal authorities, and a goodly number within the federal establishment. One of these was the curious case of the Seminole deer. In fostering Florida's cattle industry, the Department of Agriculture waged war on disease-carrying ticks, and in line with this policy it backed a drive by the state government to exterminate wild deer, which cattlemen believed to harbor ticks. Seminole Indians refused to let whites into the Big Cypress reservation to kill the deer. Harold Ickes's Interior Department, backed by its conservationist constituency, favored both the deer and the Indians. As the quarrel over the deer intensified, President Roosevelt felt obliged to intervene on the side of Interior. Suggesting that his secretary of agriculture inform the Bureau of Animal Husbandry that the deer had never been proved to be a special reservoir of disease, the president drew a political lesson. The Bureau, he remarked, had never shown that human beings were host to cattle ticks, either. "I think some human beings I know are," wrote the president. "But I do not shoot them on suspicion—though I would surely like to." In the face of high-level disapproval, the Bureau rather reluctantly gave up the fight.[25] But differences in philosophy, policy, and mission continued to divide the professional guilds and government satrapies which now crowded the resource field.

World War II interrupted a remarkable period in conservation history. The effects of the conflict were paradoxical. Though many conservation programs ended, game animals increased in number, probably because so many hunters were elsewhere, wearing uniforms and shooting at other things. The CCC was closed down (1940, the banner year for tree-planting, also saw the draft law enacted) and shortly thereafter many youths from the camps and the reserve officers who had ridden herd on them found opportunity to apply their experiences in structured communal living on a broader scale. The practical consequences of the agency's work remained, keeping its memory literally green.

Such were the peripheries of the New Deal's impact on the Depression-era South. More contradictory and far more decisive were its effects on agriculture. The strictly conservationist aspects of the New Deal program, largely embodied in Hugh Hammond Bennett's Soil Conservation Service, were fresh and informed. But other aspects of the New Deal farm policy, overall a complex and often self-confounding melange, launched a revolution in southern agriculture at grave cost to the poorest workers on the land.

Bennett himself was a son of the region, born in the North Carolina Piedmont. When he was in his twenties and working for USDA, his studies of Virginia soils convinced him of the decisive role of erosion in destroying the fertility of land. In particular he grasped the role of sheet erosion, the gradual stripping away of thin layers of soil by rain. From this perception he extended his views to form a philosophy of resource management based on the relationship between soil and cover, for erosion implied more than the familiar red gullies in southern hills. Erosion was a great elemental force, occurring naturally, yet so accelerated by human activities that a generation of two of misuse could strip off topsoil that had required millennia to form. When Bennett wrote of erosion he described a process that could transport trillions of tons of soil yearly, aggrading rivers and endangering fishlife in the process. Erosion tended to feed upon itself: topsoil, being absorbent, actually allowed less runoff than subsoils. When exposed to view, the sterile subsoil, dissected by gullies and unable to support vegetation that might impede the process, sometimes wore away to the level of rock and gravel. Certainly Bennett was not the first to discover the power of erosion, nor the first to see it as a kind of inadvertent cultural suicide, accelerated by the civilization it could help to destroy. But as a southerner, a scientist, and a publicist he was well able to feel, measure, and convey his insights.

In the bureaucracy he was both an ornament of his profession and a good bit of a nuisance. His career in the agricultural sciences took him from Alaska to the Canal Zone before depositing him in Washington, where for many years he lived the life of a government savant, with a home in the city's Northwest (later the Virginia suburbs), his job at the Department of Agriculture, evenings at the Cosmos Club. Yet he bore the reputation of a man with a bee in his bonnet, who equated erosion with the more desperate forms of sin. His writings brought him appointment in 1933 to head the newly established Soil Erosion Service in the Department of the Interior. In 1935, to the regret of some colleagues who had been glad to see him go, the bureau was shifted to Agriculture as the Soil Conservation Service. His interest in his native region remained and was reciprocated as his influence spread. He promoted the creation of state soil conservation districts as governmental subdivisions to foster erosion control and proper land use. Arkansas was the first state to enact his model legislation. The Brown Creek Soil Conservation District in his birthplace, Anson County, North Carolina, was the first district established; Alabama became the first state to be completely covered by such districts, in 1941.[26]

The heart of the philosophy he tirelessly preached was the adaptation of human endeavor to the nature of the land. His national program looked not only to erosion control by a variety of means, but to the purchase of submarginal land to take it out of production and turn it to more adaptive uses; to watershed management as an aspect of flood control; to farm forestry, drainage, and irrigation. He saw many of his ideas become law, and emerged as one of the key figures in the New Deal heyday of conservation and resource management. In 1947, in testimony before the House of Representatives, Bennett recalled his boyhood, the eroded fields of Anson County, and the fatalism of the farmers who saw the spreading gullies as "something inevitable, or a matter of course." Bennett's father "saw what was happening to his land, and he tried to do something about it; but his methods never had much of a chance to succeed.... The Bennetts were not exceptions, farmers simply didn't know how to go about stopping erosion. The necessary information wasn't there; it had not been acquired." After a career that inevitably suggests comparison with Pinchot's as expert, publicist, administrator, and mover of the men who moved the country, Bennett could reflect that that, at least, was no longer true.[27]

The same Congress that gave Bennett his first soil conservation bureau at Interior also passed the Agricultural Adjustment Act. By paying farmers not to produce and by imposing marketing restrictions on certain crops, notably cotton, the administration sought to end overproduction and raise agricultural prices, which had skidded to desperate levels. The program was one in which the South gained disproportionately, according to Brookings Institution studies. Tobacco areas profited most, with the cotton states second, while northeastern farmers appear to have benefited least. In the single year of 1933, some ten million acres were taken out of cotton production, reducing at a stroke the South's premier crop and grand staple by one-fourth. "Never again," writes an agricultural historian, "would cotton acreage return to its pre-New Deal importance."[28]

But landowners did not now require so many hands to grow crops on the reduced acreage, nor so many in ginning and handling the fiber.

Displacement of tenants had already been going on since the agricultural depression began in 1925, and had reached a peak in 1929–1932. Now it resumed as a deliberate or inadvertent consequence of government policy. In an upheaval resembling many an earlier rural exodus—the English enclosure movements or the Highland Clearances of Scotland—thousands of tenants and sharecroppers were ousted in the interest of a new and more efficient agriculture. The process went forward despite protests by the Southern Tenant Farmers' Union, by Norman Thomas, and by other defenders of the poor, and despite the efforts of the New Deal's own Farm Security Administration to make farm tenure more secure for small operators. All in all, the AAA and successor programs spawned no small amount of suffering and some of the harshest invective to emerge from a time of passionate politics.[29]

But for the landowners who received the bulk of the federal payments—and they were substantial, approaching for cotton growers alone eight-tenths of a billion dollars—a new source of income had appeared, which was soon to be supplemented by a variety of programs under the aegis of the Farm Credit Administration. For many reasons the decade 1930–1940 saw the number of southern farms shrink by 414,000 but—reversing a trend that had been underway since 1870—increase in average size, from 109 to 123 acres. Though mechanization came slowly, in response to increasing credit and the growing size of landholdings it did occur: the 4 percent of farms in the South which had tractors in 1930 almost doubled by 1940. Throughout the nation, Depression and federal programs alike worked toward elimination of small, inefficient producers. Of these the South had many, and the social effects were both harsh and necessary to any long-run improvement.

In 1941 a new impact took shape as the International Harvester Company began to produce the first commercially successful spindle-type cottonpicker. For the first time cotton farmers could harvest their staple by machinery, breaking at last the labor-intensive nature of cotton production and lessening still further the need of growers for a large, poor population of rural laborers and small producers. The onset of the national wartime emergency then opened thousands of new jobs in industry, and gave work of a different sort to the millions of men and women taken into the armed forces. The long-run result of such changes, which were to be confirmed and extended by postwar prosperity and farm policies, was a major transformation of the southern countryside by new crops, reduced rural population, and increased markets. What was on the way was nothing less than the overthrow and permanent demotion of King Cotton. By 1960 the entire South would have fewer acres in cotton than Texas alone had had in 1929.[30]

As a result, the seemingly insoluble environmental problems of the rural South were drastically altered. At grave cost to those least able to bear it, patterns of life changed which had confined too many people on a shrinking resource base, in conditions highly suitable to endemic disease. At the same time crucial technological innovations took place, partly through government action, in part simply because more people could afford them. Rural electrification brought the electric pump, which brought the flush toilet, which made hookworm transmission far less likely to occur. The collander-like cabins of the sharecroppers stood abandoned in the middle of many a field, left in islands of gums and brambles by the encircling plow, and this meant both a new kind of travail for those who had lived there, and the passing of an older kind. The houses that remained occupied were more apt to be stout and screened, with a deep well, an indoor bathroom, and inhabitants well able to afford shoes. The future promised, on average, a more prosperous people growing better adapted crops in a land where forests once again spread at the expense of cutover woodland and overcropped acres. By launching changes so extraordinary, the period of the Depression and the war earned some claim to be considered at least as revolutionary as any the South had experienced in an often turbulent past.[31]

9

South into Sunbelt

FOR SOUTHERNERS EVEN MORE THAN FOR OTHER AMERICANS, THE POST-World War II era was crowded by events. The New South system passed away as decisively if not as explosively as the Old South had a century before. Men and women who were middle-aged in the 1980s and 1990s were, like their great-grandparents, separated from their youth by more than years. The changes in the metaphorical landscape of culture were mirrored in the physical landscape, whose forms, more than ever before, were shaped by superabounding human power. Cottonfields changed to pastures or to woods; marshes to soybean fields or rolling Gulf; wild land to neon strips. As the end of the century approached, healthier, more prosperous, and more numerous southerners confronted common American dilemmas without altogether shifting the burden of a peculiar past.[1]

So quietly that their going was scarcely noticed, many of the South's endemic diseases followed the epidemic ills of cholera, yellow fever, and smallpox into practical oblivion. Pellagra had virtually disappeared by 1945. The parasitic diseases yielded to the ongoing changes in southern rural lifestyles. Hookworm disease became a curiosity. *Necator*, which had learned to thread the mazes of the human body, was undone by the suddenness of historical processes, which gave it no opportunity to adapt. The diet of most southerners became richer and more varied, and the housing of all save the very poor was improved—fundamental changes that countered disease by the positive expedient of improving health. Some pockets of extreme rural poverty did survive into the 1960s, with the old accompanying features of parasitemias and even pellagra, which had supposedly been ended; but they were rare exceptions. The pasteurizing of milk helped to control tuberculosis; though the warm, damp climate favored the disease, and its incidence remained relatively high in the cities, by 1950 tuberculosis in the South, as elsewhere in the nation, became a negligible cause of death.[2]

The enduring scourge of malaria yielded both to changes in lifestyle and to deliberate, sophisticated assault. In the South, malaria had always occupied a climatically marginal area, for the climate, though warm, was not tropical, and cool weather limited the activity of the vector during much of the year. The changing southern countryside presented many new challenges to malaria; the human reservoir decreased, houses became tighter, and the number of cattle increased to an extraordinary extent. In 1947 Congress voted funds for a National Malaria Eradication Program designed to end the disease through systematic use of the methods elaborated during the preceding decades. A double envelopment was planned: use of antimalarials to destroy the disease in its human hosts, plus vector control by DDT. In 1946 the Office of Malaria Control in War Areas became the Communicable Disease Center. Again the Surgeon General (Public Health Service) was in overall command of a campaign that, as before, depended upon cooperation with local health authorities in the endemic states. The federal government paid about three-fourths of the costs nationwide. The results were striking. In 1936, about 4,000 Americans died of malaria; in 1946, about 400; in 1952, 25; in 1965, 2—and both of those cases were from infections acquired overseas.[3]

Since, with this disease, mortality was a trustworthy index of morbidity, it was safe to conclude that *"the* Fever," as Daniel Drake had called it, had at last been defeated. Interrupting transmissions for a critical period (about three years) eliminated the parasite. Thereafter *Anopheles* might become as DDT-resistant as it chose; the vector had again become merely a nuisance of the southern dusk. In theory, the future entry into the region of even a single carrier could launch a new epidemic if the vector again became numerous and the human population, through lack of exposure, ceased to be resistant. In fact, a healthy and increasingly urban people equipped with an armory of insecticides and a modern medical system to treat the sick was well prepared to meet any new invasion.[4]

The deliverance was real, even if hard to assess in qualitative terms. What did it mean to the eight southern states that reported 181,725 cases in 1920 and almost none in 1950? Or to southern rural children, now that 5 to 20 percent of them no longer harbored the parasite? The Land of Dixie had long been the Land of Fever, its people, and especially its rural people, subject to a debilitating illness, and its reputation to the warning label that it was a sickly land, dangerous to visitors.

From these burdens, plus the enduring economic loss caused by malaria, the region was now substantially free, or at the gloomiest estimate, paroled. Yet, as with pellagra, once malaria was gone few thought much about it anymore. It was forgotten, like a dream on waking.[5]

The South has been a maker of revolutions and also the theater of revolutions made elsewhere. The fall of slavery in the 1860s and of Jim Crow a century later were no more important to the region than the fall of King Cotton. "In no major agricultural region of the United States," writes Gilbert C. Fite, "has so fundamental a change occurred. It is as if the Great Plains had abandoned wheat, or the Middle West corn, for some other crop."[6]

The overthrow of the Kingdom did not, of course, mean the end of cotton raising. In the 1970s the region still produced 80 percent of the nation's cotton, but the fiber no longer ruled. The crop that had been associated more than any other with the New South pattern gave way to a variety of successors. Commercial forests spread as corporations found in tree-farming increasing profits under favorable tax laws. As the federal government controlled cotton acreage to reduce overproduction, and synthetic fibers cut into its market, the soybean emerged as a new basic crop for the region. This Asiatic bean was a source of human and animal food and an array of industrial products—detergents, soaps, cosmetics, paints. The product of an intensely commercial agriculture, soybeans rose as cotton fell. In 1940 nine cotton-producing states from North Carolina to Louisiana turned out 7.6 million bales of cotton and 5.4 million bushels of soybeans; in 1975 the same states produced a trifle over 3 million bales of cotton, but 523 million bushels of soybeans.[7]

Crops, whether old or new, were produced by more machines and fewer hands. Between 1940 and 1974, the number of farms in the eleven southern states fell from 2.4 million to something under three-quarters of a million; during the same years, average farm size (omitting farms in Texas) rose from 86 to 235 acres. Southern farmers not only became fewer, they changed color; by the 1970s farm operators were "about 95 percent white."[8] If one seeks a revolution in contemporary America, what compares with the transformation of the black peasantry into an urban proletariat, with all the extraordinary results that followed?

On the countryside there were successor crops other than soybeans. The tung tree was introduced in 1905 and by the 1940s had become an important source of oil.[9] Rice expanded the southwestern kingdom it had established in the Gilded Age. Peanuts continued to be popular; in Georgia by the 1970s, more acres were planted to peanuts than to cotton, and the profits of the lowly goober helped to make a president. Vegetables, fruits, nuts, and a variety of other products stole acres from deposed King Cotton.

But to southerners viewing the transformation of their countryside, perhaps the most striking change was to livestock. Throughout the Black Belt that stretched from the Carolina Piedmont into Mississippi, "millions of acres of eroded cotton land were planted to grass and trees." Grasses were developed by plant scientists, and selected breeds of cattle imported to eat them. In the thirty-five years after 1940, the ten southern states east of Texas increased their cattle holdings four times over, and in terms of quality there was no comparison at all. Poultry became something of a regional obsession, and "by the mid-1950s the southern uplands were filled with . . . elongated buildings of metal, cement block, or wood sheltering hundreds or sometimes thousands of bright-feathered" fowls.[10]

Yet the South still had great cottonfields, and the boll weevil lived on. DDT had been used effectively in public health campaigns during World War II and after. It was not to be expected that the insecticide that had proved so valuable abroad against typhus-bearing lice and malaria-bearing mosquitoes would not be applied to agricultural pests at home. In the postwar years, DDT and related compounds again succeeded so well that funding for other methods of pest control dried up and entomology "became, more and more, applied chemistry." For cotton growers, the chlorinated hydrocarbon insecticides seemed to provide at last the technological fix that would end the boll weevil for good.

Heavy infestations during 1949 in the lower alluvial valley caused many growers, led by those of the delta, to adopt the new insecticides, spreading them in massive concentrations over wide areas by mechanical sprayers and airplanes. It was in this area that weevils immune to the poisons first began to appear. Weevils resistant to benzene hexachloride were noted in Louisiana in 1954, followed by the Yazoo delta, the Red River valley, and the Ouachita valley—"precisely," says an authority, "where farmers carried out the most intensive control programs." Synthetic organophosphates were then resorted to. The new pesticides appeared miraculous at first, killing weevils on contact, even killing larvae inside the buds. Reveling in the combination of mechanical and chemical powers at their disposal, growers in some parts of the South practiced "blind spraying" without regard to the level of the infestation—prophylactic spraying. "Some farmers in the coastal plain of North Carolina poisoned as many as seventeen times during the season. . . . the farmer who poisoned profited."[11]

Characterized by "persistence, mobility, and bioconcentration," the new toxins became more deadly as they moved up the food chain, reaching a peak in birds of prey. Even small quantities caused reproductive failure in Florida eagles, and the bodies of dead pelicans revealed lethal concentrations of dieldrin, endrin, and DDE, a DDT breakdown product. The impact of the toxins pervaded the environment. By killing all insects indiscriminately, broad-spectrum insecticides set off explosions of resistant pests, such as the organophosphate-resistant budworms of the Rio Grande Valley, which were freed from both competition and predation. Both the costs and the profits of insecticidal agriculture favored the big grower over the small one (though in some places small units successfully competed by hiring out the expensive spraying equipment). Well-tried management techniques were abandoned as no longer necessary. Research demands and political pressures engendered by the profits of the new agriculture brought new government interventions, supplying farmers with new poisons and also (and paradoxically) striving to suppress their most destructive results. Cotton-growing spread abroad even more rapidly than it had in the past, in large part because the synthetic insecticides were being applied there also. Competition from foreign sources, and from synthetic fibers that were the result of the same revolution in organic chemistry that had brought the pesticides, helped to cause the relative decline of cotton, bringing new demands from growers to end the career of the pesky weevil for good.

In 1958 the National Cotton Council declared war on the weevil. Extirpation was its aim. Congress appropriated $1.1 million for the struggle, and in 1962 the Boll Weevil Research Laboratory was set up near Mississippi State University. Work on a variety of control methods took place under pressure, both from the weevils, which seemed likely to develop resistance to the organophosphates as they had to the chlorinated hydrocarbons, and from a public and an environmental lobby that were increasingly hostile to insecticides. The effort to achieve complete eradication also ran afoul of deep splits among entomologists, many of whom did not believe the goal to be attainable, and of basic political problems. Who was to pay for the dubious battles of a regional interest group, and how much? Victory over the weevil, like the definitive worldwide conquest of malaria that was the object of other contemporary dreams, proved both expensive and elusive.[12]

Today a variety of control methods other than insecticides are available, but how widely such methods are used is another question. The Environmental Protection Agency (EPA) banned DDT in 1972. Resistant plant varieties are known, and the use of pheromone traps and sterile males to control the weevil population are promising. Suppression of *Anthonomus*'s numbers below the economic threshold at which returns exceed costs is possible, but the weevil lives on, held in check from year to year but not destroyed. The actual widespread use of environmentally sound programs incorporating nonchemical controls has yet to be achieved. *Anthonomus* survives, a native southerner now, the hero of many a song and story, as well it should be. It too has colonized the land, in the face of enemies armed with overwhelming cleverness and power; in nature, no more than that is demanded, and no less. Its human opponents meanwhile have encountered that law of diminishing returns that is often met with when the effort to control the environment goes too far.

Overall, the new southern agriculture exhibits a paradox that will recur in this chapter. It is incomparably better attuned to the landscape it occupies than it has been for centuries; and when the number of people it supports is weighed with its vastly improved record in regard to diversification and erosion control, its achievements are truly extraordinary. Yet it remains a major source of pollution, a danger to the variety of the landscape it has modified and helped to restore. Pesticides and chemical fertilizers used in farming form probably the greatest source of unregulated pollution in the United States today. The lack of identifiable point sources for pollution running off farmers' fields means that—from the 1898 Refuse Act to the most recent enactments on water quality—no control laws work with respect to agriculture. For the South, the national problem is worsened by the fact that the regional culture remains largely tradition-minded about pollution control, which is too often seen as an obstacle both to farmers' survival and to industrial development. The agenda for the future is evident, but for a variety of reasons, practical and political, it remains extraordinarily difficult to achieve.

Forestry, on the other hand, has advanced almost too far in sophistication. In the post–World War II years, southerners have been able to watch a continued and remarkable expansion of their woodlands. But the primeval forests have not returned. Instead, the increasing size and complexity of forest industries, the heavy investments in machinery they entail, and the long lifespans of corporations all have tended to make tree-growing an ever more important endeavor. Federal regulations and federal-state cooperation have helped to spread publicly managed woodland as well, some of it available to controlled industrial use, some held as reserves. From the U.S. Forest Service to the pulpmill, the general picture is one of longsighted management under a legion of professional foresters, some privately and some publicly employed, to serve interests that range from papermaking to backpacking.

Active tree-farming is usually traced to Weyerhaeuser's operations in the Pacific Northwest about 1940. Taken up as a crusade by the Lumber Manufacturers Association (in part at least to forestall government controls) the tree-farming movement spread rapidly during the 1940s, aided by favorable changes in federal tax laws. The basic logic of the movement was the familiar need to assure the continuity of large enterprises by securing their supply of raw material. Tree-farming brought to its climax a process reminiscent of the emergence of agriculture from primitive gathering. At first forest users merely took what they wanted; then they controlled the amount they took, reseeded, and sought to prevent damage by fire and disease. By the late 1940s intensive silviculture was in full swing.

A new staple came to the South, and with it a new monoculture. Pine woods spread that were all of a height, like well-matched soldiery, standing at equal distances, tended by professionals, and turned to calculated purposes—timber harvest and pulpmill, in the main. The South was no longer the site of a short-lived timber bonanza, as in the 1880s, but of a substantial industry growing and harvesting a renewable resource. With the Pacific Northwest, the South became one of the nation's two principal producers of wood and wood products—the national woodlots. The northwestern trees had more bulk, while the southern grew and replaced themselves more quickly.[13]

Of major importance to the character as well as the profitability of the southern forest industry was the rise of the pulp- and papermill. Mills began to dot the South during the 1920s. Over the following decades the industry learned to use hardwoods and experimented with a variety of other raw materials, including industrial wastes and agricultural residues like bagasse, the refuse of sugarcane processing. The pine tree, however, remained its foundation. Though the South produced little fine paper (chiefly, it would seem, because there had been no increase in demand since the region got into the papermaking business) it became a major producer of cheap paper and other pulp products, such as sulfate board. By the 1960s, three-fifths of domestic pulpwood was cut in the South, and, despite increased use of hardwoods, about 90 percent of southern pulpwood was yellow pine.[14]

In North Carolina, long the classic state for forests and forest industries, about 59 percent of the land area was wooded in 1937-1938. Growing stock, however, was seriously depleted, and the annual cut of softwood exceeded growth. A survey in 1956 revealed quite a new pattern. Abandonment of farmland had produced an increase of 7 percent in the area available for forest cover. There was a better balance between growth and cut, as well as a substantial absolute increase in annual growth. A 1962 survey in two coastal plain areas indicated an increase of 4 percent in the total area of commercial forestland. Over roughly the same period the area of hardwood increased relative to pine, as private owners cut pine without restocking. A local industry producing fine furniture had sprung up, taking about a quarter of the hardwood production.

In Georgia, much Piedmont land had reverted to pine as erosion, weevils, and market problems drove some owners and many tenants away. By the 1920s, attempts to destroy boll weevils by fire had combined with southern custom to make burning a genuine menace. Little remained of the estimated 36.5 million acres of virgin forest of 200 years before. Nearly 19 million acres were completely or almost completely cut over—an area of devastation probably greater than in any other state of the South. With both agriculture and woodland at a low ebb, the recovery of both began in the nadir of the Depression. In 1930 the state won an unwanted "first" by leading the nation in abandoned farms—about 65,000 that year alone. Voluntary programs of reforestation were now aided by the Civilian Conservation Corps, and the state during 1933-1934 led the South in the number of camps. In 1935 the Union Bag and Paper Company began to build the largest mill in the world. World War II stimulated reforestation by bringing high prices for timber, and the war years also saw the Forest Farmers Association founded to encourage tree-farming. In the late 1940s and 1950s the movement matured. Federal and state governments provided aid, pulp products increased, tree nurseries were set up, reforestation spread. The same years saw neighboring Alabama become the leading lumber-producing state in the South and the fourth largest in the nation.[15]

Throughout the region, state governments broke with the past to become exemplars of active aid to reforestation. The states assisted private owners through forestry commissions, or divisions of forestry within state conservation departments, and discovered an almost literally endless array of functions for state-owned forestland—as a source of wood products, as watershed protection, or for game refuges, grazing, mining, and parkways. By maintaining nurseries, giving leadership in fire control, and backing biological research into insects and the causes of disease, the states improved both their own and private holdings. They formed interstate compacts for mutual assistance against forest fires, and they received federal aid and guidance in improving their own performance and assisting private owners. As foresters learned the value of controlled burning, largely by studying southern experience and problems, science began to exploit and discipline this oldest of management practices. As forest science, business, and policy all changed together, the old migratory lumber business gave way to a permanent forest industry looking to long-term profits from a variety of products.[16]

Natural forest fared much less well in this general burgeoning of enlightened self-interest. Assaulted by developers or by urban hordes in search of recreation, areas like Florida's Big Cypress and Texas's Big Thicket were endangered by a mixed bag of loggers, builders, hunters, and campers. Hundreds of drainage and reservoir projects carried out by the U.S. Army Corps of Engineers and the Soil Conservation Service transformed stands of bottomland hardwood into reservoirs or soybean fields. Woodlands preserved without cutting in enclaves under rigorous government stewardship, on the other hand, passed via the natural forest succession into a form that, though natural, was neither varied nor particularly attractive. The steady increase of shade led to the destruction of all species except those that can sprout and grow in deepest shadow—the dark columnar glades of beech and hemlock the Europeans had found in parts of the Chesapeake region on their first arrival. The long human occupation of the South had made the typical southern woodland (if such a thing existed) at least a semimanaged forest, with some old trees, but with open meadow, shrubby edgeland, and a fire climax of longleaf pines dominating much of the lowlands and Piedmont. It was this varied woodland, which people had once played a part in creating, that too-perfect human control, either for preservation or for profit, could and did destroy.

In the commercial forests the dimensions of a great agri-industry could be seen. In 1975 there were about 1.4 million acres of monoculture pine in Louisiana alone, about 10 percent of the total forest in the state. Weyerhaeuser, Crown Zellerbach, Union Camp, Georgia-Pacific, International Paper, Scott Paper, Kimberly-Clark—the list of major companies with southern interests was a long one. In 1980 Weyerhaeuser owned 3.1 million acres in North Carolina, Mississippi, Alabama, Oklahoma, and Arkansas. Scott held 3.2 million acres spread over Washington, Alabama, Mississippi, Georgia, Florida, and South Carolina. Kimberly-Clark's 1 million acres lay mostly in Alabama, Georgia, and South Carolina. International Paper rejoiced in 4.9 million acres in the South. Georgia-Pacific grew out of a lumber company in Augusta, Georgia, moved to the west coast, but returned in 1982 to occupy a skyscraper in Atlanta. The company owned an international empire of 4.8 million acres of timberland, plus gypsum, oil, and gas. Here were new-model plantations with national corporate boardrooms for Big Houses and the sheepskin-bearing products of forestry schools for overseers.

The tree farm differs from natural forest in the types of cover and food it provides to wildlife, in its lack of genetic variety, and in its lesser total biomass. Phosphate fertilizers and an occasional dose of herbicide stimulate growth and remove competitors. Productivity increases greatly, reaching multiples of the natural forest growth in an equal span of time. Careful selection and culling ensure the survival of healthy trees. The tree farm's look and feel differ from those of any other woodland. Visitors will never find there the forest giant, nor will they experience the mysterious and moving complex of forms that is shaped by spontaneous growth. Pine plantations look disconcertingly like cornfields in Brobdingnag.

While some mills engrossed large areas of forestland, others found greater economy and better public relations in buying from many small producers. Some processors encouraged the culling of pinelots, taking warped and damaged trees and enabling small owners to improve their stand by systematic thinning; others worked through agents who all too often did the reverse. Pulpmill payrolls caused bankers and merchants to discover the importance of forestry, and private landowners who supplied the mills became more sophisticated as time went by, giving up their worst trees and keeping the best. On the whole, the pulp industry developed as an environmental plus at its alimentary end. The same could not be said of its highly toxic effluents.

Economic pressure drove the South to develop and made its state governments more than normally complaisant about the consequences to the landscape. Southern politicians, reacting as they had since the Civil War to the hope of industrial progress, enacted some of the weakest pollution control standards in the nation. "When a paper mill was constructed in a little Alabama town some 20 miles from Montgomery," wrote Marshall Frady in 1970, "it tinged even the Capitol's corridors, on especially muggy mornings, with a vague reek: [George] Wallace, who was then Governor, would note on such mornings, 'Yeah, that's the smell of prosperity. She does smell sweet, don't it?' "[17]

Yet the many and complex faces of reforestation in the South can be neither easily summarized nor dismissed in a phrase. Its influence may already have reached cosmic proportions, as signaled by some recent evidence that, after centuries as a net producer of carbon dioxide and other carbon by-products of burning, the United States east of the Mississippi and south of the Ohio is now a carbon sink, absorbing more of the element than it emits. In a world seemingly bent on cutting down its forests and roasting itself by the greenhouse effect as carbon dioxide increases in the atmosphere, this development, if confirmed, is not unwelcome. As southern agriculture becomes more profitable, however, and as cleared land continues to rise in value, fields have once again begun to advance at the expense of forests. A new swing of the pendulum has started with as yet uncertain long-range effects.[18]

The wildlife that woodlands and marshes shelter has come to depend more than ever before on the position of individual species in a world dominated by humans. Many species, especially those that are highly specialized or of restricted range, face extinction. Florida, with its shrinking marshlands and rapidly swelling population, stands third on the list of states in the number of its endangered species, and the Southeast has in the past century been the site of about a quarter of all

American extinctions. Among southern animals on the official list of endangered and threatened species maintained by the Interior Department are the Florida panther, the Everglades kite, the Mississippi sandhill crane, the ivory-billed woodpecker (which may be extinct already), and the whooping crane, plus an array of little-known creatures, including fish, amphibians, snails, and a long roster of mussels. The manatee and the sea turtle eke out a precarious existence. By the end of the 1980s, Florida alone held 39 endangered species, 26 that were deemed threatened and 45 others that were listed as "of special concern."[19]

Yet never did the urge to preserve wildlife seem so strong. A list of comeback species could be compiled as easily as a list of endangered ones and would include, both nationally and in the South, the beaver, white-tailed deer, egret, heron, wild turkey, and wood duck. In the Florida keys, "crowded and noisy breeding colonies" of sooty terns, once devastated by eggers and plume-hunters, increased from a low of about 5,000 adults in 1903 to about 100,000 in the 1960s. In Louisiana, state wildlife and fisheries experts successfully reintroduced the brown pelican during the 1970s, after pesticides had annihilated it in the state it symbolizes. The alligator recovered under protection to the point that controlled hunting could begin again. The same was true of the beaver. Despite renewed taking, both species continued to grow under "harvesting," the beaver so much as to become "responsible for millions of dollars worth of damage to merchantable timber and crops." Crop damage by deer also increased throughout the Southeast despite the legal killing of almost one million animals a year. The system of conservation agencies, legal controls, federal and state refuges, game management areas, public forests, and parks has achieved an enormous reversal for dozens if not hundreds of species whose fate, three generations earlier, seemed either dubious or sealed.[20]

The likelihood that any species can continue to survive, however, depends as never before upon human interest or sentiment. Species live or die out depending on whether they inhabit a human-favored and expanding habitat like the forest or a declining and abused one like the marshes. Even within the forest, survival may depend on whether a species fits into the new managed woodland. The relationship of deer to tree farms provides a remarkable case of unplanned convergence between the practices of industry and the needs of wildlife. Edgelands between cut and uncut areas and the practice of controlled burning cause the spread of plants favored by the deer for browsing, and make the tree farm a favorite habitat. Species needing a climax forest are in a less happy situation, and beaver, for evident reasons, are definitely not wanted. Polluted streams and marshes can endanger even terrestrial wildlife; in 1980 alarmed game management officials found mercury contamination in the Florida panther, one of "our most endangered mammals," apparently because the panthers were eating raccoons that in turn had eaten contaminated fish and crustaceans. Among game species, waterfowl may be the most susceptible to pollutants, including both heavy metals from industrial processes and agricultural pesticides. Add in declining habitat in the wetlands, and their future looks bleak indeed.[21]

Management practices in park, forest, and farm cut across the needs of wildlife in paradoxical ways. Controlled burning, though good for the deer, eliminated undergrowth and dead trees, depriving some non-game birds of habitat. Yet in areas where the forest has been a fire climax for millennia, fire control programs can be destructive to other birds. In commercially grown woodlands, almost anything that abates the regularity of the tree farm and increases patchiness and variability—the occurrence of natural sloughs where pines will not grow, the use of helicopters for seeding instead of seeding by hand—is a plus for wildlife. So is the burning of undergrowth every four or five years instead of annually. Different kinds of vegetation standing at different heights, and varying food sources, are needed to maximize survival.

Public relations also makes a great difference in determining which species live on and which do not. Some are favored by a spectacular or merely cute appearance, while obscure or homely creatures pass unwept. The peregrine falcon finds partisans that the dusky sparrow lacks; some creatures that are threatened shelter themselves in bureaucratic niches on the endangered species list, while others, no less in peril, are excluded by changes in the winds of national policy. State wildlife agencies grow ever more professional, but their orientation is overwhelmingly toward the interests of hunters, fishermen, and trappers. To be favored by sportsmen and their pressure groups is an ambiguous advantage to individual animals and birds, but one that makes for species survival. And the opposite is true also. Lifestyle changes among southerners, a growing repugnance for bloodsport, and competition from a variety of other outdoor activities caused the number of hunters in the region to remain static or decline during the 1980s, despite the expanding population. Wildlife managers saw in the trend "less public support for hunting and a corresponding decrease in political support for wildlife management programs."[22]

As in the past, some species prosper by drawing close to human activities, by serving as recreation or food, or by entering ecological niches that people themselves have created. Raccoons, opossums, and squirrels have always adapted to semiagricultural and even urban environments. About 1948 an African native, the cattle egret, entered Florida by way of South America. In the decade that followed, the egret became common in several southern states, finding around the growing cattle industry an opportunity to thrive. And there are species that spread on their own with small regard for human wants or objections. The armadillo has steadily expanded its range in the Southwest, crossed the Mississippi, and moved as far east as Florida. The fire ant, after entering the country via the port of Mobile in 1918, proved so successful as to provoke in the 1970s a sharp confrontation between environmentalists and the USDA, southern states, and farmers engaged in spraying dieldrin more or less at random in hopes of destroying it. In 1976 the EPA banned the pesticide as a carcinogen; both before and after the banning, the march of the fire ant continued.

The most subtle of impacts upon wildlife derived from the transformation of real wilderness into a kind of garden or recreational preserve. Constantly growing human power reshaped nature so extensively that even preserves of wild land became theaters of a human thrill—the "wilderness experience." Even here the wild world that once had existed independent of humanity and indifferent to its power had largely become extinct.

In the postwar era, fishlife continued to find a substantial degree of protection in the human appetite for food and sport. Fish farms made catfish and crayfish into domesticates. In 1949 the fishing-tackle industry organized the Sport Fishing Institute, whose programs of information, research, and lobbying at the national and state levels were aimed, not unnaturally, at the goal of improving sport fishing. Marine research advanced under the patronage of southern universities; the Chesapeake Biological Laboratory was set up by the University of Maryland in 1925, the Duke University Marine Laboratory in 1937, and the University of Miami's Institute of Marine Science in 1943. Both pure and applied research have been of significance to the commercial fisheries, about half of whose production is currently drawn from the nation's Atlantic and Gulf coasts.[23]

Yet by the 1990s overfishing and pollution had seriously depleted the resource that so many were trying to save. The Chesapeake suffered from industry, urbanization, commerce, and pesticides, with intermittently disastrous effects on its fishlife and the centuries-old folk culture that fishing, crabbing, and oystering supported. In the Gulf of Mexico similar impacts followed swiftly. As the century approached its end, the runoff of pesticides and fertilizers polluting the Mississippi had produced a Chesapeake-sized seasonal dead zone near the river's mouth, where oxygen depletion caused by algal blooms and subsequent decay made fishlife impossible. A growing national taste for seafood, an increasing population, and a fad for Cajun cooking also contributed to what an expert called "strip mining the Gulf Coast waters."[24] How long the fisheries can survive under siege from burgeoning use and declining habitat is unclear, but their condition at this writing is clearly perilous.

Urged on by a sense of danger and impending loss, during the 1960s and 1970s a new upsurge of conservationist feeling ran an extraordinary course in the United States. The environmental movement, born of postwar changes in national outlook that emphasized the quality of life rather than the more elemental demands of survival, underwent hothouse growth in the 1960s. The upheavals of the decade—a time of grave social stress—created demands for new priorities in almost every aspect of national life. The need for an environmental cleanup was debated early and often, and pioneering legislation passed under President Lyndon Johnson.

But the period of most intense concern followed the election of Richard Nixon and the end of the Vietnam War. Then the movement exploded into an impassioned crusade as militant youth found a new outlet in "eco-tactics." Environmentalism, however, was too broad to be defined by a single motive or a single generation. Its issues provided a series of media events but evinced too much strength to be dismissed as a fad. Assailed by ideological enemies in the 1980s, it survived to influence—though never again to dominate—the agenda of both major parties.[25]

Despite the growth of active environmentalist groups throughout the South, and despite the strong pro-environmental stands of Georgian Jimmy Carter and Tennessean Albert Gore, the movement met strong opposition in the region. In its opposition to development, environmentalism differed strikingly both from the Progressives and the New Dealers who had gone before it. While earlier modes of conservation had furthered the organization of America and consequently possessed a positive economic appeal, environmentalism at the grassroots level often took the form of a citizen guerrilla war waged to impede the actions of corporations and governments alike. A diffusion of expertise among the people during the preceding generation now raised up lawyers, scientists, and self-taught enthusiasts to dispute the in-house experts of great corporations on their own ground—aided by federal laws, often with success.

One result was a transformation of values, by which many of the achievements of New Deal resource policy came to be seen as insults offered to nature by human presumption. Another, in the South, was the hostility of business and political leaders whose views now tended to parallel those of the West rather than, as in times past, those of the Northeast. Significant was the new view of the Tennessee Valley Authority held by many environmentalists, and the curious history of the snail darter and Tellico Dam, which revealed much about the temper of the time and helped to prepare a political eclipse of the environmental movement.

The evolution of TVA from radical experiment to pillar of the regional establishment has already been noted. In the late 1970s one critic termed TVA "an agency which began life playing kickball with the liberals" but "finally waddled off to the golf course with the conservatives." Crucial to the change in image was its role as a power producer. During World War II TVA turned to steam generation after using up most of the hydropower potential of the river valley. During the Eisenhower administration the agency survived a major attack by private power interests (to date, its last fight for life). In the years that followed, TVA's success as a power company spawned new controversies and dangers. It became the nation's largest purchaser of coal, it contracted for stripmined coal, and its coal-fired steam plants became major sources of sulfur dioxide pollution. Such expansion was demanded by local power users, who saw in the agency a cheap road to prosperity.[26]

TVA began to acquire a whole new set of enemies: farmers who claimed its air pollutants were defoliating their crops, landowners forced out of its reservoir and recreation sites, and even coal producers who had lost out in the bidding. By maintaining its 1953 rate structure into the energy-conscious 1970s, TVA sold power most cheaply to large users and overestimated the growth of demand. Meantime it went into nuclear power in a big way. Not only among enthusiasts, but among many segments of the public as well, TVA was in a fair way to winning itself the image of an unresponsive bureaucracy serving itself and its clients with small regard for the environment or energy conservation. In 1977 the EPA joined the states of Alabama and Kentucky, as well as a number of environmental groups, in bringing suit against the agency for its chronic violation of federal clean air laws.

Local opposition, aided by national environmental groups, had also formed to a number of TVA projects, including the Tellico Dam, planned for the Little Tennessee River. A $124 million multipurpose development, the project was to impound a 33-mile stretch of the region's last free-flowing river. Its reservoir would cover 16,500 of the 38,000 acres of the river valley, providing flood control and recreation but very little additional power, and that only by supplying additional water to the nearby Fort Loudoun Dam on the Tennessee River via a canal. TVA would gain less than one-tenth of one percent in additional generating capacity from the project—23 megawatts added to an existing capacity of 27,000. Construction began in 1967.

The valley of the Little Tennessee meant different things to different people. The Boeing Corporation intended to develop 21,500 acres of residential, commercial, and industrial property there. Local farmers hoped to keep their land. Cherokee Indians saw a onetime center where their culture had flourished before the Removal. Archeologists saw important sites. Fishermen saw good trout fishing. Writer and mystic Peter Matthiesen saw "in the soft light of early November, the muted fire colors of the fall . . . [and] the clear, swift water rolling down from the ridges of the Great Smoky Mountains, to the east." In 1973 a University of Tennessee zoologist discovered one of the obscurest attractions of the valley, a "three-inch, tannish, bottom-dwelling member of the perch family," which at the time was believed to live nowhere else. The snail darter thereupon acquired a Latin name—*Percina tanasi*—and a footnote in history.[27]

Four months after the snail darter's discovery. Congress passed a law requiring federal agencies to ensure that actions they carried out should not jeopardize endangered species, and instructing the secretary of the interior to prepare a list of such species. The new law fatefully converged with the purposes of those who, for quite different reasons, wished to stop the dam. Local citizens and the Environmental Defense Fund had already sued to stop Tellico on the grounds that TVA had failed to provide an environmental impact statement, delaying the project for a year and a half. When the statement was prepared and approved by the court, Tellico's opponents turned elsewhere. In 1975, over TVA's objections, the snail darter was placed on the list of endangered species because of the threat posed by Tellico. A new suit by a Tennessee law professor led in 1977 to a decision by the Sixth Circuit Court of Appeals enjoining completion of the dam, the judges ruling that "conscientious enforcement of the Act requires that it be taken to its logical extreme." On June 15, 1978, the Supreme Court upheld the Court of Appeals. The fact that TVA had transplanted 700 snail darters to the Hiwassee River was not viewed by the courts as important; the question was one of congressional authority and the procedures established by the law. The Court of Appeals, however, acknowledged that "TVA has not acted in bad faith. Its efforts to preserve the snail darter appear to be reasonable."

The image of a dam stopped by a "minnow" proved irresistible to journalists. Less happy was its effect on the Endangered Species Act and the national environmental movement, whose organizations had tried without success to avoid fighting on the snail darter issue. The act was being used against a project that proponents had so far failed to defend rationally, or opponents to stop on its fairly conspicuous lack of merit. A federal committee set up to consider the issue refused to exempt Tellico from the act. Tennessee lawmakers, including the Republican leader in the Senate, Howard Baker, reacted to the impasse by attaching a rider to an appropriation bill that exempted the project from the provisions of the Endangered Species Act—and, for good measure, all other acts as well. President Jimmy Carter reluctantly signed the bill with the Tellico rider. A last and ironic point came when the snail darter's original discoverer found a colony of the fish that TVA had transplanted to the Hiwassee River before the Tellico floodgates closed, "thriving and reproducing." Moreover, he found another colony in a creek eighty miles from the Tellico Dam, which apparently had been there for years.[28]

The final note of the controversy rang in August 1980 when the Cherokees failed to appeal their own suit against the dam, which had lost in a lower court. Beyond its strictly local and temporary result—and its telling nationwide impact on the environmental movement—the controversy had revealed in striking fashion how far both American conservationists and TVA had come from their beginnings.[29]

The South's appetite for federal projects did not become dulled by feeding. Few projects were more earnestly desired or longer sought than the Tennessee-Tombigbee Waterway, a bold design to connect the two rivers in order to create a north-south water highway paralleling the Mississippi.

Found to be uneconomic during the New Deal heyday of great water projects, Tenn-Tom won congressional authorization in 1970, just as the environmental crusade was reaching full intensity. Construction under the requirements of federal environmental laws resulted in a project that in all likelihood has been more thoroughly studied than any other in the often stormy history of such endeavors. Large and vulnerable to criticisms on many grounds—chiefly environmental and economic—the waterway faced legal attacks by the Louisville and Nashville Railroad and the Environmental Defense Fund. The Corps of Engineers was accused of faking the cost-benefit ratio to make Tenn-Tom appear legally feasible and of introducing modifications into the project that exceeded its congressional mandate. For a time construction halted on Tenn-Tom's midsection after the Fifth Circuit Court of Appeals ruled that the engineers had "blatantly violated" the National Environmental Policy Act and their own regulations by failing to file an amended environmental impact statement covering the project changes. Congressional support eroded, yet remained sufficient to maintain funding.

Politics was, in fact, the key to the project's ultimate success. President Richard Nixon provided support as he pursued his southern strategy; President Carter sought to extract political benefit from water projects he had tried but failed to defeat; black leaders sought to use their growing clout to provide jobs and contracts for minorities; and southern congressmen single-mindedly pursued economic development through federal subsidy. The end result was a waterway that provided markedly fewer economic benefits than enthusiasts had hoped—larger tows (barge trains) moved faster on the Mississippi than on the lock-impeded waterway—and that may have marked the end of an era in federal water projects. After Tenn-Tom, growing national suspicion of big government and the pressures of a mounting federal deficit helped to bring in laws and regulations that required greater local contributions and increased user fees, while the Corps of Engineers turned much of its energies to projects aimed at restoring the environment.[30]

While federal projects faced growing opposition from environmentalists, regional industrialization contributed to a steady increase in population and wealth that strengthened southern influence on national affairs to a level not seen since before the Civil War. Well into the 1970s, industrial development in the South continued to rest heavily, as it traditionally had, upon textiles and raw materials. The importance of southern mineral resources grew steadily, as advancing technology discovered new uses for them and new ways of extracting and processing them. Arkansas mined some 97 percent of the nation's domestic bauxite, while Florida phosphate supplied 70 percent of the national total. North Carolina provided much of the nation's lithium, a light metal used for a fantastic variety of purposes from controlling manic-depressive disorders to cleaning suburban swimming pools. Clays were put to many uses; manganese and copper were extracted in eastern Tennessee and the Carolinas.[31]

Industry often kept a distinctive regional form, favoring small towns or the countryside rather than forming vast agglomerations around the cities. Typical was the North Carolina pattern, with its dispersed, nonurban manufacturing, that at one point in the 1960s enabled the state to be simultaneously the nation's fifteenth most industrialized and its seventh most rural. The automotive industry followed a similar pattern during its extraordinary invasion of the South. Nissan built a $1.2-billion plant in Smyrna, Tennessee, employing some 6,000 workers; General Motors a $3-billion plant for its Saturn line in Spring Hill, Tennessee; Toyota a $2-billion plant in Georgetown, Kentucky; while BMW and Mercedes-Benz invested almost as heavily in South Carolina and Alabama, respectively. Yet southern cities burgeoned as well, drawing electronics and aerospace companies, among many others, and making great strides in the management of information and money. Atlanta became a giant of the communications industry, and by the mid-1990s Charlotte, North Carolina, ranked as the country's third largest banking center.[32]

But progress frequently came at a price. States fought for the manufacturer's dollar by granting tax breaks and other favors, whether the citizens could afford them or not. Both old and new industries produced effluents, and some posed threats to human health. Black lung caused by coal dust in mines and byssinosis caused by lint in spinning mills were problems of long standing. But many pollutants belonged to the brave new world of modern chemistry. Industrial and pesticide pollution mingled in a ghastly episode at Hopewell, Virginia, where workers at the Life Science Products Company, a firm under contract to Allied Chemical Corporation, were poisoned by Kepone, a chlorinated hydrocarbon. After several were found to be suffering from anxiety, tremors, inability to focus their vision, and low sperm counts, federal and state inspectors entered the plant—a former gasoline service station with two outbuildings and a paved area known to workers as the "pad"—and found extraordinary conditions.

Workers were covered with a "fine white Kepone dust," wet Kepone flowed into the pad area, no protective clothing or boots had been provided, and employees often ate lunch in the production area. Effluents had entered sewers, storm drains, and a nearby creek that fed the James River. The plant was closed in August 1975. In December, after high levels of Kepone turned up in fish and shellfish, the state of Virginia forbade fishing in the James River. Impaired human beings, multimillion-dollar lawsuits, and a temporarily wrecked fishing industry were among the consequences of a disaster to which utter corporate irresponsibility and belated action by state and federal officials both contributed.[33]

Of far greater potential impact was the growing petrochemical industry. Oil wells spread, not only over much of the southwestern landscape but also into the waters of the Gulf of Mexico, where the Martian shapes of the oil rigs gave an added unreality to misty mornings in those rich fishing grounds. Along the hundred-mile corridor of the Mississippi River from New Orleans to Baton Rouge, brilliant floodlights lit up industrial sites filled with metal spheres, cylinders, and endless pipes. The industry, located among poor and heavily black communities where jobs often seemed more important than health, spawned conditions that to African American activists suggested a new and lethal form of discrimination. Benefiting from the same political complaisance that papermill owners found elsewhere in the South, the petrochemical industry joined upstream companies and agriculture in pouring massive discharges of pollutants into the Mississippi River, the air, and the land—until the 1970s, with utter lack of restraint.

Then in 1973 Dow Chemical, viewing with alarm the possibility that 273 million pounds of toxic liquids and sludge it had previously dumped might enter the Plaquemines Acquifer, began drilling the first of hundreds of wells to pump out and treat the effluent. In the next decade, a $1-billion lawsuit brought against Georgia Pacific and Georgia Gulf by Reveilletown, Louisiana, alleging damage to both health and property, resulted in a secret but apparently significant out-of-court settlement. Dow Chemical then bought out a hamlet called Morrisonville and moved its people to new homes, in a gesture that might be viewed as either industrial statesmanship or fear of future litigation. Fines levied by the EPA and state regulators provided another incentive for change. Between 1988 and 1992, some impressive reductions were reported by Louisiana's Department of Environmental Quality. Yet the state remained first in the nation in toxic emissions from petrochemical industries, and legal loopholes at both the national and the state level left many types of dangerous waste unreported. The effects on public health remained in doubt; cancer death rates along the industrial corridor were abnormally high and increasing. But the precise causes were, as usual, hard to pin down; occupational exposure to carcinogens formed only one element in the lifestyles of workers and their families.[34]

The environmental costs of fossil fuel production centered on the stripmining of coal and on oil spills in coastal waters. Stripping was not, of course, confined to coal; phosphates were mined in the same way, but in areas where the land was flat. Coal's broad distribution and mountainous locations combined with the flow of often toxic water from the mining sites to make the problems raised by open mining acute. Stripmining control bills aimed in part at compelling the companies to restore the natural contours of the landscape were twice vetoed by President Gerald Ford before a third act was passed and signed by President Carter in 1977. In 1980 Congress enacted legislation to list, examine, and ultimately clean up the most serious sites where hazardous wastes of all types had been discharged. In the mid-1990s, about a quarter of the 1,082 General Superfund sites were in the old Confederacy plus Kentucky and West Virginia, as well as a third of the 150 federal facilities that had also contributed to the nation's pollution problems. The General Superfund sites were a rogue's gallery of pesticide and chemical dumps, municipal landfills, mine sludge disposal areas, and oil pits. Florida had the most—54 sites in all—from a combination of real problems and a high level of environmental activism.[35]

Petroleum in the South is recovered primarily from two belts, one inland, the other running along the western Gulf Coast and extending offshore. Typically the inland fields are large—the East Texas field in the late 1960s held more than 25,000 wells—while those in the coastal region are "small but numerous and highly productive." Natural gas, a fuel notably clean-burning and efficient, is the Cinderella of the fields; once "flared" as a nuisance, it has become over the course of forty years a national fuel of increasing importance. By the 1970s gas was being sought deliberately to be fed into the national system of distribution by pipelines.[36] The offshore wells, combined with the movement of supertankers and the loading and offloading of crude oil, created increasing environmental hazards to coastal ecosystems.

The popular view of oil spills tended to be based on television images of gunky brown stains on blue water, volunteers cleaning oilsoaked shorebirds, and wildlife dead of hypothermia and starvation. In fact, spills on the open sea appeared to have few long-term effects. But in marshlands the results could be severe, sometimes devastating. Here crude oil slowly releases its aromatic hydrocarbons, the ring molecules like benzene that are toxic and hard to degrade. Tidal oscillations and the wind might wash a single spill repeatedly over such an area. Shellfish were found to be particularly susceptible, and even oysters and mussels that survived a spill might fail to reproduce the next spring. The distribution of plankton was also affected, implying a disturbance at the very base of the food chain. Studies at the Patuxent Wildlife Research Center in Maryland showed that waterfowl embryos might be poisoned inside the egg, and that the toxins could impair the growth of flight feathers and damage the livers and kidneys of fowl.[37]

In recent decades both wells and tankers have suffered disasters that polluted the South's coastal waters. Offshore production began in Louisiana in 1947, and the rigs when operating normally were accepted as natural entities by local marine life; underwater growths formed on the structures, and fish learned to hide from enemies and pursue prey in and around them. In turn, fishermen learned to hover around the rigs, and despite conservationist alarms the sport of rig fishing gained both a name and an extensive following. The structures proved to be volatile neighbors, however. The early 1970s saw a number of incidents in the Gulf as Chevron, Shell, and Amoco rigs caught fire.

But the most striking spill in southern waters—indeed the largest in history to date—came with the blowout of a Mexican well, Ixtoc I, that the government-owned Petroleos Mexicanos had drilled in the Gulf off Yucatan.

For several months crude oil escaped at the rate of more than a million gallons a day. The oil was picked up by a gyre—"a far-flung, slow-moving whorl of ocean currents, revolving in a clockwise direction." The gyre spun the oil northward, where seasonal winds began to push it toward the Texas shoreline. Here one of the nation's most remarkable coastal areas lay in its path, "a combined aviary, zoo, aquarium, and arboretum stretching 350 miles and immensely more variegated than any man-made preserves." Small rivers had carried sediment into the Gulf, dropping the material as their velocity slackened to form bars that wind and currents then shaped into barrier islands. Quiet lagoons lay within the barrier, with wetlands beyond. Along the coast were seven national refuges and a dozen National Audubon Society refuges, all alive with hundreds of species of plants, animals, and birds. A busy recreation industry, tourism, commercial fishing, and Corpus Christi's $200 million-a-year shrimp fleet also depended on the area. Led by the Coast Guard and advised by the

National Oceanic and Atmospheric Administration, five federal agencies held the line against the oil slick at the outer shore of the barrier islands until a seasonal windshift in September carried the spill offshore, where it could be broken up and dispersed by wind and waves. The victory really depended on the weather; strong winds could have carried the spill into the lagoons, and a hurricane would have brought disaster.[38]

Since the industry is likely to increase before it declines, more spills may be expected in the future. Importation of foreign oil poses evident dangers, while the 1989 Exxon Valdez disaster in Alaska served as a reminder that the domestic oil industry sometimes moves its products in the same way and with potentially the same effects. Moreover, the discharge of oily ballast and leakage from small tankers can produce a chronic release of toxins no less dangerous in the long run than the spectacular spills. Since society appears unlikely to give up the search for new sources of oil, or production from the old, a considerable burden has been thrown upon the companies and the government to exert utmost care in minimizing the dangers, if only to avoid a political or public-relations cataclysm. Increased research and improved techniques and organization in managing spills appear to be the best hope for decades to come. But because so many incalculable factors are at work, the distinctive natural and human communities of the Texas and Louisiana coasts—to name only the most threatened areas—will remain at risk for the foreseeable future.[39]

Oil pollution formed only a small part of the complex story of the southern coasts and wetlands, however. From the Chesapeake to the Rio Grande, the coasts were over-rich in natural and human activities, and they did not always coexist in harmony.

The shore is a region of evanescent forms, shaped by transient balances between forces of erosion and those of deposition. Since the sea level appears to be rising along much of the southern coast, erosion has an evident advantage, irrespective of people's actions. But some human endeavors have helped to tip the balances in favor of erosion or, by halting it in one place, intensified it in another. Favored structures to protect the shore include groins, which stand at right angles to the shore, and bulkheads and seawalls, which parallel it. Groins are supposed to cause deposition, while seawalls separate the land and water with the aim of preserving the former. The problems caused by such structures are many, but two may be cited. By obstructing the littoral drift, groins build up one area while starving another of sand needed to compensate for erosion. Bulkheads interrupt the biological functioning of shoreline areas, especially in natural wetlands, which demand the mingling of land and water. Artificial channels, human destruction of vegetation on barrier islands, and even levee systems, though far inland, may adversely affect the shore. Such changes in the landscape prevent the natural overflow of rivers, impede land-building, and weaken the coast's protection against storms.

In the South, major cities impinge directly on the shoreline from the Chesapeake to Galveston, while the beaches, from the Eastern Shore of Maryland and Virginia to Corpus Christi, Texas, lure a horde of vacationers and sun-starved settlers. By 1980 three of every five Americans were said to live within fifty miles of salt water. The popularity of the shores impelled the development of the country's most transient real estate, raised property values, and brought in government programs of construction, subsidized flood and disaster insurance, and the like. As investments rose, so did demands for protection. Yet never was an endeavor so literally built on sand—even more so in the South than elsewhere, because hurricanes are so frequent.

On the Atlantic shore, the beach resorts evolved during the twentieth century from empty dunes and scattered bungalow-and-hotel enclaves patronized by the wealthy to crowded playgrounds of the masses. The rise in oceanfront property values led to demands that the shifting of the sands be halted—Canute American style—and created divisions within oceanside communities between beach-living summer visitors and inland-dwelling year-round residents. Yet many forms of development contributed to the transience that people hoped to stabilize. Coastal recession increased after the 1950s, as the clearing and leveling associated with construction lowered the profile of barrier islands, allowing even moderate storm tides to wash over them. Studies repeatedly showed that control methods were both temporary and "site specific"—what worked in one area failed in another. The National Park Service, steward of the national seashores, tended during the 1970s to turn away from artificial methods of nourishing and protecting beaches with pumped sand and control structures. But governments, like the shore itself, were at the mercy of the public, and private beaches and common seashores alike were subjected to contradictory policies demanded by the visitors who came to enjoy nature and, once arrived, loved it to death.[40]

In south Florida development went farthest. Although the state's eastern cities lay upon the coastal ridge, communities also pushed out onto the barrier islands, bestowing the full panoply of development— walls, groins, construction of high-rise buildings—upon the Gold Coast. Severe and repeated storms then caused not only shoreline recession but scenes of devastation (notably during Hurricane Andrew in August, 1992) resembling the disasters of war.

Near the tip of the peninsula, urban development is compressed into a comparatively narrow space along with highly productive agriculture and the unique Everglades. The Everglades does not resemble more conventional swamps dynamically, for instead of being a natural sump like the Okefenokee (formed around a lagoon cut off from the ocean by a coastal ridge) its natural form is a "river of grass" periodically overflowed by Lake Okeechobee and stoppered at its southern end by coastal mangrove thickets. The rhythmic alternation of wet and dry seasons, the rich diversity of wildlife, and the Big Cypress to the northwest of the Everglades combined to form a region without parallel. Attracted by the rich soil, Floridians early embarked on efforts to drain the marsh and block the overland flow of water by dikes. Meanwhile, a major city took form along the Atlantic coast.

For the Everglades, Miami could have been a worse neighbor. The city was not a major commercial port, had no petrochemical complex as yet, and occupied itself with recreation, retirement, light industry (electronics, wood products, boats), and the drug traffic. Of greater impact were corporation farms south of Lake Okeechobee (notably those of the profitable and heavily subsidized sugar industry), which demanded ever more water, returning it to the Everglades heavily polluted, if they returned it at all. The rising population recruited from the North and Cuba—and indeed from everywhere—combined with commercial agriculture and the landform and weather to create that ultimate paradox, a bizarre struggle for water in one of the nation's wettest regions. Between 1948 and 1971 elaborate structures were built by the Corps of Engineers to distribute the waters of Lake Okeechobee among the city, agriculture, and the Everglades. The latter was often the loser.[41]

Florida, unique in the size of its population of relative newcomers, benefitted by their energy and fresh ideas in both its politics and its environmental movement. Few southern states could show so able and productive a group as the Florida scientists, politicians, and plain folk who stopped the Cross-Florida Canal in 1969 and forced the resiting of the Miami jetport in 1972. The reverse side of the "Florida Experience"—as Luther J. Carter named his exhaustive account of the state and its problems—was an evolution that in many ways removed it, and especially its southern half, from the South and created an all-Sunbelt suburbia without traditions or a sense of place. The problems facing wildlife grew with the population; the Game and Freshwater Fish Commission routinely provided its officers with survival training, which they needed as apprehending fugitives and finding corpses began to enliven their customary round of duties. Speeders endangered panther crossings on the Tamiami Trail and dumpers used wildlife sanctuaries as trash bins. One violator, apparently the employee of a mobster, was apprehended throwing 2,500 liters of lethal toxins into the Everglades. "Needless to say," wrote the Commission's James A.

Ries, "he never offered to implicate his boss," despite fines totaling $4.6 millon.[42]

By the 1980s the reduction of the Everglades to about half its original size, the diversion of water from what remained, and phosphorus pollution from agriculture had created serious dangers to the ecosystem, signaled by declining populations of waterbirds and an intermittent dead zone offshore in Florida Bay. In 1988 the federal government sued the state, demanding that Florida obey its own water-quality laws. Three years later the state mandated a water improvement and management plan for the Everglades. The next year, Congress authorized the Corps of Engineers to restore the once meandering Kissimmee River, converted by the agency in earlier days to a canal romantically designated C-38, and to develop a master plan for revising the region's hydrology. The usual donnybrook of special interests accompanied the development of the multipurpose plan without concealing its importance. If the region can actually be, as the *New York Times* put it, "replumb[ed]" in such a way as to revive the Everglades without destroying agriculture or denying the Gold Coast water, a major construction project of the past will have been reversed successfully for environmental purposes. Such an achievement might mean a new beginning, not only for the Everglades but also for the state and the federal agencies that first endangered the region and by 1994 stood poised to join hands in the work of restoration.[43]

The Gulf coast has its own problems. Fine sand beaches, neon strips, storm-harried barrier islands, and intermittent marshes formed by alluvial rivers reach almost to the border of Louisiana, a state that contains roughly 40 percent of the nation's coastal wetlands. Then the waters of the Mississippi River enter, both through its own delta and through that of its largest distributary, the Atchafalaya River, a hundred miles to the west. The river system drains about two-thirds of the forty-eight contiguous states, builds wetlands that provide habitat for millions of migratory birds, and empties the pollutants of at least eleven states into one of the nation's most important commercial fisheries. New Orleans lies upriver from the Gulf roughly a hundred miles; the Intracoastal Waterway cuts across the marshes, and artificial channels have been dug, or natural ones straightened and expanded, to move water traffic to the port or to service the oil rigs, or both.

While many factors shape the coast—currents and winds prominent among them—the balance of forces demands the deposition of new sediment to counter natural subsidence and the rising of the sea. The overflows of the Mississippi built both of the existing deltas as well as four ancient ones. Land-building in the marshes depends on the erosion of the deltas by coastal currents that distribute the soil westward. Because the balance between deposition and erosion is at best a fine one, human activities have disturbed it in a number of ways. The Mississippi levees and their extensions, the jetties at the mouths of the passes, have channeled the alluvium down the continental slope into deep water, reversing the land-building process, and Louisiana at present may be losing more land area than any other state. New ponds appear in the marsh and gradually enlarge to the size of small lakes, or, breaking into the Gulf, form irregular embayments. As the coast retreats, it loses the ability to trap sediment, accelerating the process, and the salt water of the Gulf moves inland through spreading embayments and canals, destroying vegetation whose roots help to bind the soil. With the land vanishes both wildlife habitat and a buffer against hurricanes. By 2040 the expanding west-bank suburbs of New Orleans may lie on or close to the shoreline of an embayment of the Gulf of Mexico—a dismaying prospect, considering the land's low profile and the frequency of hurricanes.[44]

By contrast, the delta of the 130-mile-long Atchafalaya is growing, fed by the diversion of silt-bearing water from the Mississippi, of which it may one day become the main channel. (Some studies cast doubt on the ability of control structures built by the Corps of Engineers at the Atchafalaya's junction with the Mississippi to prevent diversion indefinitely.) High water on the Mississippi during the 1970s accelerated the growth of the delta; apparently by 2050 about a hundred square miles of new wetlands will have been created. In the meantime the interior of the valley fills as well, transforming portions of a vast and once-primitive marsh into arable land and setting the stage for struggles among local farmers, fishermen, environmental groups, recreation interests, and oil and gas companies—all refereed by the Corps and the state of Louisiana—to define the future of the nation's largest freshwater swamp, assuming that it has one.[45]

The debate over the wetlands of Florida and Louisiana underlines the fact that the Corps of Engineers, once the devil of the environmental movement, has seen its power over rivers and wetlands expand continuously with the movement's success in enacting pro-environmental laws. The permit power originally granted the agency to protect the navigability of the nation's waterways has been extended—through the federal National Environmental Policy Act (1969), the Clean Water Act (1977), and a number of court decisions—both to nonnavigable waters and to environmental purposes. As a result, "by the mid-1970s, the Corps found itself courted by some of its staunchest critics in the environmental community." Still hostile to its civil works mission, environmentalists support its regulatory function, while within the Corps itself a new generation of engineers seeks new sources of funding in undoing the projects of generations past. And not only the Corps has changed. Environmentalism too has been transformed from a rebellious theory to an enduring fact, embodied in the very establishment its enthusiasts once railed against.[46]

How much the new mission of the Corps will mean to the vanishing Louisiana coastline is another question. Although the loss of wetland has slowed over the past decade from about forty to about twenty-five square miles a year, that is still substantial. As a Corps spokesman remarked, "It is ironic that the [levee] system which brings prosperity and security to humans is literally costing them the earth beneath their feet." A recent law sponsored by Senator John Breaux has made available small but steady funds—$30 to $40 million a year—for local restoration projects, with the Corps chairing an effort that involves seven federal and four state agencies. The projects have been likened to "a Band-Aid approach," however, while Corps proposals for major restoration, now on the drawing board, will require billions that may not be available.[47]

Seen from the Gulf, the southern coastline presented itself to the first explorers as a distant line drawn through the horizon haze. On the Louisiana coast, the line has since become dotted, and by the middle of the next century will lie in some areas 35 miles to the north of its ancient location. Elsewhere, human activities have changed the land; here they have interacted with natural forces to cause the land itself to disappear slowly beneath the advancing sea.

Such in brief are a few of the milestones that marked the transformation of the southern environment in the postwar years. It would be quite easy, by taking a carefully selective view, to show that the overwhelming display of human power typical of the time created either a garden or a briar patch. Enduring threats like malaria ceased to matter, forests spread over once-eroded land, crops were adapted to the landscape with unwonted skill, selected breeds of cattle were introduced to graze on selected grasses. Though disappearing habitats endanger many species, never in the region's history has so much land, or so much skill and care, been devoted to the perpetuation of wild animals. Yet the burden of industrial and agricultural pollution has grown constantly, much of it unregulated and unrecorded. Today a kind of autointoxication threatens the health of humans, animals, birds, and fish, even as the old diseases vanish.

Pogo the Possum once summed up the dark side of the era in his celebrated epigram, "We have met the enemy, and he is us." It is worth remembering that after many victories, there is now nobody else left to fight. As if enacting some ancient fable, humankind, having slain many dragons, is obliged to confront itself. As the South increases in population, wealth and power, it needs more than ever men and women of humility and knowledge.

10

Myth and Dream: An Epilogue

FROM THE FIRST ADVERTISERS OF THE WONDERS OF VIRGINIA, SOUTHERNERS have displayed the normal human urge to brag about their region. Sometimes their fellow Americans, entranced by the image of a warm place they have really known very little about, have joined them in undiscriminating praise—whenever they were not doing the opposite. A decade ago historian Charles P. Roland delivered a witty address on "The South, America's Will-o'-the Wisp Eden," pointing to the persistent promise and lagging performance of the region, which he interpreted in terms of its enduring educational backwardness and economic colonialism. The urge to overcelebrate the promise has had many sources. Seventeenth-century pamphlets praised a transatlantic Eden while settlers died like flies. In the 1850s, Hinton Rowan Helper constructed an image of a South rich in all natural blessings and held back from attaining its promise solely by the curse of slavery. This notion, which might be termed the Protestant Legend—if you enslave your fellow man, God in His anger will make you poor—has proven to have enormous staying power. Unfortunately, its two basic elements, the suppositions that slavery is uneconomic and that the South's natural resources are remarkably rich, appear to have no basis in reality. In the bootstrapping of the New South crowd, in the "homicidal . . . wowserism" that H.L. Mencken discerned in the 1920s, in the long effort of historians to find *the* villain of southern history that makes it peculiar and un-American, the significance of a modest and much-abused environmental dower has been pretty consistently undervalued.[1]

Rhetoric has generally been easier. Of the staple crops of the South, has any grown more lush? "This land, this South," wrote William Faulkner, "for which God has done so much, with woods for game and streams for fish and deep rich soil for seed and lush springs to sprout it and long summers to mature it and short mild winters for men and animals."[2] To the booster the Eden lies always ahead, waiting development; to the romantic, behind, never more to be attained. Meanwhile southerners dwelt for generations in a land in which hard labor, including the hard labor of thought, proved barely sufficient to sustain life.

Myth can signify either an illusion or an insight. The South has yet to produce its Thoreau, but it has not lacked for a diffuse and oracular poetry of the wild (and more commonly of the rural). With few and mostly recent exceptions, southern writers have apprehended the landscape as country people do to whom animals, trees, and landforms are not to be named only but encountered. They have often written of these other beings with understanding and humor, both rooted in a deep intimacy, like people telling anecdotes about family members to other family members.

To Marjorie Kinnan Rawlings in rural Florida, the moccasin that does not strike her skittish horse, though almost stepped on, is a gentleman; she raises her hat to him when safely past. Through Cross Creek the farm animals wander at will in the casual southern way, living their lives at roadside.

> A lean sow was being serviced by a boar. The young of her previous litter seized the moment of her immobility to nurse. I spoke of it to Norton, mentioning that the sow had a thoughtful aspect.
> "Of course she was thoughtful," he said. "She was thinking, 'It's a vicious circle.'"

The author's city-raised cat goes blissfully wild, her highly urban terrier disdains the woods. Sandspurs, in the malicious way of their kind, strike straight for her case of poison ivy. At the end of her chronicle, too, rhetoric flowers: "Cross Creek belongs to the wind and the rain, to the sun and the seasons, to the cosmic secrecy of seed, and beyond all, to time."[3] Seasons, seed, and time make the rural world, and its chronicler admits no human being save in that context.

There are some high moments in the apprehension of the southern landscape in the works of Mark Twain and his odd contemporary, Lafcadio Hearn. Hearn was an outsider, an anadromous fish who swam the oceans over without finding the stream of his birth. Much of the oddity of *Chita* derives from the fact that it has only one memorable character, the Gulf plain, which is never a mere setting. Rather a slim story of wraithlike people makes a conventional setting for the land. The realities are the Mississippi River, flowing in August "like a torrent of liquid wax," the Greek chorus of billions of frogs, the relentless fevers, the pitiless sun, and the great upwellings of the sea.[4] If nature here absorbs the human, on Huck Finn's Mississippi something much rarer happens. In a seemingly effortless *tour de force* the river becomes a giant metaphor—of escape, freedom, natural communion—while always remaining its own real, closely observed, and native self. This recalls those Japanese prints in which a branch or wind-jostled flower forms a definitive element of design while it retains a perfect, almost photographic realism. One must accept the demonstration that such things can be; it seems a waste of time to try to understand them.

Among more recent masters, Faulkner inevitably draws the eye of anyone seeking a nature poet. Nature and man's relation to it run through most of his work, and the thematic outrage that shapes its strange forms derives not only from man's misuse of man, but from his misuse of the southern land. The compassionate Isaac McCaslin carries much of the burden of both forms of human sin in *Go Down, Moses*. As a youngster entering the deep woods in "a slow drizzle of November rain just above the ice point," he brings to the wild something that responds to it and recognizes it—the other and inner wild reborn in every human child and only slowly and imperfectly cut back, trimmed and pruned and made ultimately into a strange topiary, the civilized man. Here too is peace. Not bent to human purpose, the wild knows nothing of purpose frustrated; because it is the working together of myriad things, each following its own nature, it is in fact one. So the story of the bear concerns not only the compulsion of the hunt that "void[s] all regrets and brook[s] no quarter," but the ultimately discovered identity of slayer and slain.[5]

If there is a Thoreau of the modern South, he may be the Mississippi painter and solitary Walter Inglis Anderson. Before his death in 1965 he painted thousands of watercolors, many showing the creatures and plants of Horn Island, a sandy barrier off the Mississippi Gulf coast to which he withdrew with greater frequency as he grew older. He also filled some eighty or ninety logs with his daily observations, the changes of the seasons, his sharply observed but transient contacts with people, and notes on his paintings. Art was not, in itself, the object of his withdrawal from the common life, a withdrawal that left his wife with four children to support as best she could. Painting, like Horn Island, was Anderson's road to a union with nature. The union was unsentimental, observed with sharpness and humor by the man who both experienced and portrayed it. His lack of illusion was perhaps related to the mosquitoes, which, he noted, would bore through a rock to get at him. He was not awed by sunsets, remarking that many were "simply wild explosions of color" designed to "stun people into a state of mute wonder."

Indeed, he was apparently not very interested in what is called scenery at all. The island of dunes and sloughs had its moments—splendid when a sunset was "arranged with taste," awesome when a hurricane beset it—but what mattered about it was that it was the abode of life. Anderson painted life, not scenes. "The mangroves are terribly vital," he wrote, with their sprouting of "pulpy tuberous forms" among which were the holes of the fiddler crabs. He watched the crabs at sundown "moving their bodies either to the right or left, then raising their large claw," and suddenly saw himself through their eyes, "some huge presence . . . between them and the golden disc of the sun." Out of a life increasingly separated from men and woven of such experiences he defined himself against conventional reality:

Man begins by saying, "of course," before any of his senses have a chance to come to his aid with wonder and surprise. The result is that he dies and his neighbors and his friends murmur with the wind, "of course."

The love of bird or shell which might have restored his life flies away carried by the same wind which has destroyed him. The bird flies and in that fraction of a fraction of a second man and the bird are real. He is not only King he is man, he is not only man he is the only man and that is the only bird and every feather, every part of the pattern of its feathers is real and he, man, exists and he is almost as wonderful as the thing he sees.[6]

The religious and philosophical roots of conservation have been much explored since Aldo Leopold called almost fifty years ago for a new ethic embracing all of nature. There are many possible sources for such an ethic—the Christian doctrine of stewardship, the Calvinist abhorrence of waste, the elemental instinct of any generation to think of its children's needs as well as its own. The historian Joseph Petulla has argued for the emergence of an ethic through the very process of rulemaking: the compromises of today will become the norms of tomorrow. Such suggestions are just and reasonable. Yet there remains in the thought of the movement's leaders from John Muir forward a wildwood note of natural mysticism that transcends the practical issues, the quotidian evolution of custom and law, and the discoveries of science. It did not seem right to end this essay without taking some note of southern artists who have brought back from a country more mysterious than the deep woods the personal vision of a waste and wild divinity that binds all tangible forms.[7]

Since mankind first entered the hemisphere, skirting the Pleistocene glaciers, he has done many foolish and cruel things to the land and the creatures on which his own life depends. Much hard-learned good sense has also been in evidence, in Indian myth and custom, white and black science and law, and everyman's poetry. The wisdom which has sustained the human journey may yet prove to be the goal of it, as well.

Notes

Epigraph (p. v): *The American Environment: Readings in the History of Conservation* (Reading, Mass., 1968), ix.

Introduction

1. J.H. Parry, *The Spanish Seaborne Empire* (New York, 1966), 65.
2. Quote from Frederick A. Cook, Larry D. Brous, and Jack E. Oliver, "The Southern Appalachians and the Growth of Continents," *Scientific American* 243 (Oct. 1980): 163. See also Charles B. Hunt, *Physiography of the United States* (San Francisco, 1969), 166; H.H. Read, *Geology: An Introduction to Earth-History* (London, 1949, 1963), 169-70.
3. Hunt, *Physiography*, 93, 155-56; Joseph M. Petulla, *American Environmental History: The Exploitation and Conservation of Natural Resources* (San Francisco, 1977), 23-24.
4. Hunt, *Physiography*, 91; H.B. Vanderford, *Managing Southern Soils* (New York, 1957), 23.
5. Quote from Francis Butler Simkins, *A History of the South* (New York, 1953), 593. See also Avery Odelle Craven, *Soil Exhaustion as a Factor in the Agricultural History of Virginia and Maryland, 1606-1860* (Gloucester, Mass., 1965), 12-19.
6. Quotes from Hunt, *Physiography*, 96, 150-51; and Lafcadio Hearn, *Chita* (New York, 1889), 15. See also Charlton W. Teabeau, *Man in the Everglades: Two Thousand Years of Human History in the Everglades National Park* (n.p., 1968), 26-31; Isaac Monroe Cline, *Tropical Cyclones* (New York, 1926), 24-25; Robert W. Hamson, *Alluvial Empire* (Little Rock, Ark., 1961), 43.
7. Henry Rose Carter, *Yellow Fever: An Epidemiological and Historical Study of Its Place of Origin* (Baltimore, 1931), 14-16, 30-32; Mark Boyd, ed., *Malariology: A Comprehensive Survey* . . . (Philadelphia, 1949), 15. See also chapter 2 for further information on these diseases.
8. P.B. Medawar and J.S. Medawar, *The Life Science: Current Ideas on Biology* (New York, 1977), 48, 54; Stephen Jay Gould, "The Ghost of Protagoras," *New York Review of Books* 27 (Jan. 22, 1981): 44.
9. Quotes from Ulrich B. Phillips, *Life and Labor in the Old South* (Boston, 1929), 3; and William H. Sears, "The Southeastern United States," in Jesse D. Jennings and Edward Norbeck, eds., *Prehistoric Man in the New World* (Chicago, 1964), 259. A recent and excellent analysis of the influence of southern climate on the region's culture is A. Cash

Koeniger, "Climate and Southern Distinctiveness," *Journal of Southern History* 54 (Feb. 1988): 21-44.

Chapter 1

1. Charles Hudson, *The Southeastern Indians* (n.p., 1976), 37; Gordon R. Willey, *An Introduction to American Archeology: North and Middle America* (Englewood Cliffs, N.J., 1966), 48, 61-62.
2. Willey, *Introduction*, 44.
3. Quote from Charles Darwin, *The Voyage of the Beagle*, ed. Leonard Engel (Garden City, N.Y., 1962), 401. See also P.S. Martin and H.E. Wright, eds., *Pleistocene Extinctions: The Search for a Cause* (New Haven, Conn., 1967); Ernest L. Lendelius, Jr., "Vertebrate Palaeontology of the Pleistocene: An Overview," in R.C. West and W.G. Haag, eds., *Ecology of the Pleistocene: A Symposium* (Baton Rouge, 1976), 55ff; Hudson, *Southeastern Indians*, 41.
4. Willey, *Introduction*, 31, 48-49; Hudson, *Southeastern Indians*, 42, 45-47.
5. Willey, *Introduction*, 62, 265.
6. Ibid., 267; Hudson, *Southeastern Indians*, 57-59, 61-63.
7. George E. Stuart, "Mounds: Riddles from the American Past," *National Geographic* 142 (Dec. 1972): 795-96.
8. Quote from George W. Beadle, "The Ancestry of Corn," *Scientific American* 242 (Jan. 1980): 112. See also Walter C. Galinat, "The Origin of Corn," in G.F. Sprague, ed., *Corn and Corn Improvement* (Madison, Wis., 1977), 1-48.
9. Quotes from Beadle, "Ancestry," 112; and John Lawson, *History of North Carolina* (London, 1709), 76.
10. Hudson, *Southeastern Indians*, 94-97; Kenneth F. Kiple and Virginia H. Kiple, "Black Tongue and Black Men: Pellagra and Slavery in the Antebellum South, " *Journal of Southern History* 43 (Aug. 1977): 411.
11. Lawson, *History of North Carolina*, 43, 68; Peter P. Cooper, "The Southeastern Archeological Area Re-Defined," *Quarterly Bulletin of the Archeological Society of Virginia* 26 (1972): 136-37; Philip Alexander Bruce, *Economic History of Virginia in the Seventeenth Century* (New York, 1935), 1:78, 88n. For an older and sharply contrasting view of the Indian use of fire, see Hu Maxwell, "The Use and Abuse of Forests by the Virginia Indians," *William and Mary Quarterly*, 1st ser., 19 (Oct. 1910): 73-103.
12. Quotes from Kenneth H. Garren, "Effects of Fire on Vegetation of the Southeastern United States," *Botanical Review* 9 (Nov. 1943): 617-18; and Hunt, *Physiography*, 114. See also I.F. Eldredge, "Administrative Problems in Fire Control in the Longleaf-Slash Pine Region of the South," *Journal of Forestry* 33 (1935): 342-46; B.W. Wells, "Ecological Problems of the Southeastern United States Coastal Plain," *Botanical Review* 8 (1942): 533-61. Gordon M. Day, "The Indian as an Ecological Factor in the Northeastern Forest," *Ecology* 34 (1953): 329-46, presents solid evidence of a similar effect in the Northeast.
13. Quote from Bruce, *Economic History*, 1:175.
14. Quotes from Pierre de Charlevoix, *Journal of a Voyage to North-America*... (London, 1761), 2:107; and William Byrd, *London Diary and Other Writings* (New York, 1958), 573. See also Gary B. Nash, *Red, White and Black: The Peoples of Early America* (Englewood Cliffs, N.J., 1974), 283-85. On energy requirements, see Gerald M. Ward, Thomas M. Sutherland, and Jean M. Sutherland, "Animals as an Energy Source in Third World Agriculture," *Science* 208 (May 9, 1980): 570.
15. Hudson, *Southeastern Indians*, 159, 272. On the Indian as natural ecologist, see the enlightening discussion in Calvin Martin, *Keepers of the Game* (Berkeley, 1978), 157-88.
16. David B. Quinn, *North America from Earliest Discovery to First Settlements: The Norse Voyages to 1612* (New York, 1977), 6-7.

17. William B. Rye, ed., *The Discovery and Conquest of Terra Florida by Don Fernando de Soto* . . . (New York, 1966), 92.
18. Frederick W. Hodge, ed., *Spanish Explorers in the Southern United States, 1528-1543* (New York, 1907), 30-31, 68, 80.
19. Quinn, *North America*, 564.
20. Quoted in Wesley F. Craven, *The Southern Colonies in the Seventeenth Century* (Baton Rouge, 1949), 44.
21. See Paul H. Hulton and D.B. Quinn, *The American Drawings of John White*, 2 vols. (London, 1964).
22. Quotes from Craven, *Southern Colonies*, 22; Richard Hakluyt, *Hakluyt's Collection of the Early Voyages, Travels, and Discoveries of the English Nation* (London, 1810), 3:330-31; and Thomas Harriot, *A Briefe and True Report of the New Found Land of Virginia* . . . (Frankfurt, 1590), 16.
23. Edmund S. Morgan, "John White and the Sarsaparilla," *William and Mary Quarterly*, 3d ser., 14 (July 1957): 414-17; Quinn, *North America*, 448; St. Julian R. Childs, *Malaria and Colonization in the Carolina Low Country, 1526-1696* (Baltimore, 1940): 83-86. Ironically, *cinchona*, a genuine specific for malaria, was little esteemed by the Indians, who apparently did not have malaria until after the conquest, and was not introduced into Europe until 1640. Boyd, *Malariology*, 4.
24. Hakluyt, *Collection*, 3:302.
25. Henry F. Dobyns, "Estimating Aboriginal American Population: An Appraisal of Techniques with a New Hemispheric Estimate," *Current Anthropology* 7 (1966): 395, 441-42. See also idem, *Their Number Became Thinned: Native American Population Dynamics in Eastern North America* (Knoxville, Tenn., 1983); and the discussion in Russell Thornton, Jonathan Warren, and Tim Miller, "Depopulation in the Southeast after 1492," in John W. Verano and Douglas H. Uberlaker, *Disease and Demography in the Americas* (Washington, D.C., and London, 1992), 187-96.
26. Willey, *Introduction*, 309-10. On the possible American origin of syphilis, see E.G. Wakefield and Samuel C. Dillinger, "Diseases of Prehistoric Americans of South Central United States," *Ciba Symposia* 2 (1940): 458-60; Saul Jarcho, "Some Observations on Diseases in Prehistoric America," *Bulletin of the History of Medicine* 38 (1964); and idem, *Human Paleopathology* (New Haven, Conn., 1966), 91-97; Alfred W. Crosby, *The Columbian Exchange: Biological and Cultural Consequences of 1492* (Westport, Conn., 1972), 122-64; Benjamin L. Gordon, *Medicine throughout Antiquity* (Philadelphia, 1949), 405-6; C.W. Goff, "Syphilis," in Don Brothwell and A.T. Sandison, eds., *Diseases in Antiquity* (Springfield, Mass., 1967), 279-93; C.J. Hackett, "The Human Treponematoses," ibid., 152-69.
27. T.D. Stewart, "A Physical Anthropologist's View of the Peopling of the New World," *Southwestern Journal of Anthropology* 16 (Autumn 1960): 264-65; Childs, *Malaria and Colonization*, 25-26.
28. William H. McNeill, *Plagues and Peoples* (Garden City, N.Y., 1976), 132-75.
29. Quote from Wilcomb E. Washburn, *The Indian in America* (New York, 1975), 105-7; Crosby, *Columbian Exchange*, 39. See also Hudson, *Southeastern Indians*, 105.
30. Quotes from Rye, *Discovery and Conquest*, 56-58. See also Hudson, *Southeastern Indians*, 110.
31. Quote from Crosby, *Columbian Exchange*, 40. See also Henry Savage, Jr., *Discovering America, 1700-1875* (New York, 1979), 3; Willey, *Introduction*, 252, 307-8; David B. Quinn, *The Roanoke Voyages* (Cambridge, Mass., 1955), 1:378.
32. Hakluyt, *Collection*, 3:311, 318, 340.
33. Ibid., 338.
34. Quinn, *North America*, 566.
35. Crosby, *Columbian Exchange*, 74, 88; Craven, *Southern Colonies*, 19.

Chapter 2

1. Carville V. Earle, "Environment, Disease, and Mortality in Early Virginia," in Thad W. Tate and David L. Ammerman, *The Chesapeake in the Seventeenth Century: Essays on Anglo-American Society* (Chapel Hill, 1979), 96-125; Gordon W. Jones, "The First Epidemic in English America," *Virginia Magazine of History and Biography* 71 (Jan. 1963): 3-10; Lyon Gardiner Tyler, ed., *Narratives of Early Virginia, 1606-1625* (New York, 1907), 127-28. Karen Ordahl Kupperman, "Apathy and Death in Early Jamestown," *Journal of American History* 66 (June 1979): 24-40, presents evidence for a psychological dimension to the crisis.

2. Abbot Emerson Smith, *Colonists in Bondage: White Servitude and Convict Labor in America, 1607-1776* (Gloucester, Mass., 1965), 125-26, 215-18.

3. Quote from ibid., 215. See also John Duffy, "The Passage to the Colonies," *Mississippi Valley Historical Review* 38 (June 1951): 21-38; Wesley F. Craven, *White, Red, and Black: The Seventeenth Century Virginian* (Charlottesville, Va., 1971), 3; Bruce, *Economic History*, 1:135-38, 595-96, 600-601, 612, 625-27. On the transportation of children for labor, see Peter Wilson Coldham, "The 'Spiriting' of London Children to Virginia, 1648-1685," *Virginia Magazine of History and Biography* 83 (July 1975): 280-87.

4. Darrett B. Rutman and Anita N. Rutman, "Of Agues and Fevers: Malaria in the Early Chesapeake," *William and Mary Quarterly* 33 (1976): 31-60; Wyndham B. Blanton, *Medicine in Virginia in the Seventeenth Century* (Richmond, 1930): 50-55; Boyd, *Malariology*, 9. See also Keith Dahlberg, "Medical Care of Cambodian Refugees," *Journal of the American Medical Association* 243 (March 14, 1980): 1064.

5. John Duffy, *Epidemics in Colonial America* (Baton Rouge, 1953), 10, 13-14; Craven, *Southern Colonies*, 147; Wyndham B. Blanton, "Epidemics, Real and Imaginary, and Other Factors Influencing Seventeenth Century Virginia's Population," *Bulletin of the History of Medicine* 31 (Sept.-Oct. 1957): 454; Smith, *Colonists in Bondage*, 254.

6. Craven, *Southern Colonies*, 96; Bruce, *Economic History*, 1:138, 626. On Bermuda, see Wesley F. Craven, *An Introduction to the History of Bermuda* (n.p., n.d.).

7. Paragraphs based on: Bruce, *Economic History*, 1:143-44, 498-99; Francis Jennings, *The Invasion of America: Indians, Colonization, and the Cant of Conquest* (Chapel Hill, 1975), 27; Blanton, *Medicine in Virginia*, 35; Thomas B. Robertson, "An Indian King's Will," *Virginia Magazine of History and Biography* 36 (1928) 192-93; Craven, *White, Red, and Black*, 59-60; John Reed Swanton, *Indian Tribes of North America* (Washington, D.C., 1911), 71; Byrd, *London Diary*, 573; Daniel B. Smith, "Mortality and Family in the Colonial Chesapeake," *Journal of Interdisciplinary History* 8 (Winter 1978): 403-27; Alfred W. Crosby, "Virgin Soil Epidemics as a Factor in the Aboriginal Depopulation in America," *William and Mary Quarterly* 33 (Apr. 1976): 289-99.

8. Peter A. Fritzell, "The Wilderness and the Garden: Metaphors for the American Landscape," *Forest History* 12 (Apr. 1968): 17; Samuel R. Aiken, "The New-Found Land Perceived: An Exploration of Environmental Attitudes in Colonial British America" (Ph.D. diss., Pennsylvania State Univ., 1971), 154.

9. Quotes from John Smith, "Description of Virginia," in Tyler, *Narratives*, 90; and Nash, *Red, White, and Black*, 314. See also Bruce, *Economic History*, 1:157; Ralph H. Brown, *Historical Geography of the United States* (New York, 1948),19; John Smith, *Generall Historie of Virginia, New-England, and the Summer Isles* (London, 1624), 147. The voice of the promoter echoes again in Lawson's endorsement of North Carolina; see, for example, *History of North Carolina*, 43.

10. Smith, "Description of Virginia," in Tyler, *Narratives*, 90; Hunt, *Physiography*, 15759: Craven, *Southern Colonies in the Seventeenth Century*, 105-7.

11. Craven, *Southern Colonies in the Seventeenth Century*, 115.

12. Quote from Craven, *Soil Exhaustion*, 61. See also John E. Pomfret, with Floyd M. Shumway, *Founding the American Colonies, 1583-1660* (New York, 1970), 35; Robert Beverly, *The History... of Virginia*, ed. Louis B. Wright (Chapel Hill, 1947), 57-58. Patents issued

under headright to 1650 averaged 446 acres; in 1650-1700, 674 acres. Bruce, *Economic History*, 1:527-32. On tobacco, see J.E. McMurtrey, Jr., *Distinctive Effects of the Deficiency of Certain Essential Elements on the Growth of Tobacco Plants in Solution Cultures*, USDA Technical Bulletin No. 340 (Washington, D.C., 1933), 19.

13. Craven, *Soil Exhaustion*, 32. For an account of the process of cultivation (though at a later date), see Joseph Clarke Robert, *The Tobacco Kingdom: Plantation, Market, and Factory in Virginia and North Carolina, 1800-1860* (Gloucester, Mass., 1965), 32-50.

14. See Harriot, *Briefe and True Report*, 14: "The ground they [the Indians] never fatten with mucke, dounge, or any other thing; neither plow nor digge it as we in England."

15. Lawson, *History of North Carolina*, 76; Harold B. Gill, Jr., "Wheat Culture in Colonial Virginia," *Agricultural History* 52 (July 1978): 380-93. Quote from Carville V. Earle, *The Evolution of a Tidewater Settlement System: All Hallows Parish, Maryland, 1650-1783* (Chicago, 1975), 18.

16. Paragraphs based on: Bruce, *Economic History*, 1:201, 302; 2:6-8; Craven, *White, Red, and Black*, 26; Allan L. Kulikoff, "Tobacco and Slaves: Population, Economy and Society in Eighteenth-Century Prince George's County, Md." (Ph.D. diss., Brandeis University, 1976), 5-6, 26, 41, 63.

17. Quote from Brown, *Historical Geography*, 20-21, 62. See also Louis Cecil Gray, *History of Agriculture in the Southern United States to 1860* ([1932]; reprint, Gloucester, Mass., 1958), 1:42-45.

18. Henry H. Merrens, *Colonial North Carolina in the Eighteenth Century: A Study in Historical Geography* (Chapel Hill, 1964), 89-90, 173-81; Lawson, *History of North Carolina*, 75.

19. Quoted in Merrens, *Colonial North Carolina*, 119.

20. Gray, *History of Agriculture*, 1:57-59; Mark Catesby, *The Natural History of Carolina, Florida, and the Bahama Islands* (London, 1731-1748; reprint, Savannah, Ga., 1974), 13-14.

21. Merrens, *Colonial North Carolina*, 13; Brown, *Historical Geography*, 39.

22. Johann Martin Bolzius, "Reliable Answer to Some Submitted Questions Concerning the Land Carolina, in Which Answer, However, Regard Is Also Paid to the Condition of the Colony of Georgia," trans. and ed. by Klaus G. Loewald, Beverly Starika, and Paul S. Taylor, *William and Mary Quarterly*, 3d ser., 14 (Apr. 1957): 226. Bolzius's reports have recently appeared in book form: George F. Jones and Don Savelle, eds., *Detailed Reports on the Salzburger Emigrants Who Settled in America* . . . (Athens, Ga., 1983).

23. William McKee Evans, "From the Land of Canaan to the Land of Guinea: The Strange Odyssey of the Sons of Ham," *American Historical Review* 85 (Feb. 1980): 354; Winthrop D. Jordan, *White over Black: American Attitudes toward the Negro, 1550-1812* (Chapel Hill, 1968), 63, 71-72, 74-76.

24. Philip D. Curtin, "Epidemiology and the Slave Trade," *Political Science Quarterly* 83 (June 1968): 198-200; Craven, *White, Red, and Black*, 77-78; Evans, "Land of Canaan," 39-40; Childs, *Malaria and Colonization*, 25; Gordon Harrison, *Mosquitoes, Malaria and Man: A History of the Hostilities since 1880* (New York, 1978): 5n; Dennis G. Carlson, "African Fever, Prophylactic Quinine, and Statistical Analysis: Factors in the European Penetration of a Hostile West African Environment," *Bulletin of the History of Medicine* 51 (Fall 1977): 386-87; Todd L. Savitt, *Medicine and Slavery: The Diseases and Health Care of Blacks in Antebellum Virginia* (Urbana, Ill., 1978), 27-35.

25. Quote from Bruce, *Economic History*, 2.82. On the African disease environment, see "The Disease Factor: An Introductory Overview," in Gerald W. Hartwig and K. David Patterson, eds., *Disease in African History: An Introductory Survey and Case Histories* (Durham, N.C., 1978), 3-24.

26. Lawson was told by Indians in North Carolina that the nose-destroying illness was native. See Lawson, *History of North Carolina*, 14; Todd L. Savitt, "Filariasis in the United States," *Journal of the History of Medicine and Allied Sciences* 32 (Apr. 1977): 140-50; Thomas C. Parramore, "The 'Country Distemper' in Colonial North Carolina," *North Carolina*

Historical Review 48 (Jan. 1971): 44-52. Parramore emphasizes the difficulties in distinguishing among the different treponematoses. See also Carter, *Yellow Fever*, 81-83; Duffy, *Epidemics in Colonial America*, 207; Richard H. Shryock, *Medicine and Society in America, 1660-1860* (New York, 1960), 85; David Geggus, "Yellow Fever in the 1790s: The British Army in Occupied Saint Domingue," *Medical History* 23 (Jan. 1979): 38-42; John Hunter, *Observations on Diseases of the Army in Jamaica* (London, 1796), 233-51; Andrew J. Warren, "Landmarks in the Conquest of Yellow Fever," in George K. Strode, ed., *Yellow Fever* (New York, 1951), 6; Thomas C. Parramore, "Non-Venereal Treponematoses in Colonial North America," *Bulletin of the History of Medicine* 44 (Nov.-Dec. 1970): 571-81.

27. Ira Berlin, "Time, Space, and the Evolution of Afro-American Society on British Mainland North America," *American Historical Review* 85 (Feb. 1980): 47; Shryock, *Medicine and Society*, 83. Quotes from Jordan, *White over Black*, 262; and Brown, *Historical Geography*, 131. See also JoAnn Carrigan, "Privilege, Prejudice, and the Strangers' Disease in Nineteenth Century New Orleans," *Journal of Southern History* 36 (1970): 578; and Savitt, *Medicine and Slavery*, 39-41.

28. Paragraphs based on: Berlin, "Time, Space, and the Evolution," 71, 73; Kulikoff, "Tobacco and Slaves," 64, 89; Joseph I. Waring, *A History of Medicine in South Carolina, 1670-1825* ([Charleston ?], 1964), 35. Quotes from Edmund Berkeley and Dorothy S. Berkeley, *Dr. Alexander Garden of Charles Town* (Chapel Hill, 1969), 59, 124. See also Marion M. Torchia, "Tuberculosis among American Negroes: Medical Research on a Racial Disease, 1830-1950," *Journal of the History of Medicine and Allied Sciences* 32 (July 1977): 252-80; Savitt, *Medicine and Slavery*, 33; Joseph C. Miller, "Mortality in the Atlantic Slave Trade: Statistical Evidence on Causality," *Journal of Interdisciplinary History* 11 (Winter 1981): 385-424.

29. Quote from Berlin, "Time, Space, and the Evolution," 61. See also James M. Clifton, "Golden Grains of White: Rice Planting on the Lower Cape Fear," *North Carolina Historical Review* 50 (Oct. 1973): 365-86; Childs, *Malaria and Colonization*, 126-28, 146-47, 188, 219; Waring, *History of Medicine*, 10-12, 23; Shryock, *Medicine and Society*, 86. An interesting study of mortality in a colonial port city, emphasizing the role of immigrant ships, is Billy G. Smith, "Death and Life in a Colonial Immigrant City: A Demographic Analysis of Philadelphia," *Journal of Economic History* 37 (Dec. 1977): 863-89.

30. Quote from Bolzius, "Reliable Answer," 240.

31. For a brief account of the plasmodium's life cycle, see Harrison, *Mosquitoes, Malaria and Man*, 114-20; on the habits of *A. quadrimaculatus*, see Stanley J. Carpenter and Walter J. LaCasse, *Mosquitoes of North America (North of Mexico)* (Berkeley, 1955), 52-53.

32. John Duffy, "Smallpox and the Indians in the American Colonies," *Bulletin of the History of Medicine* 25 (1951): 335, 338-39; Berkeley and Berkeley, *Dr. Alexander Garden*, 136-38. Quotes from John R. Swanton, *Indians of the Southeastern United States* (Washington, D.C., 1946), 104, 111, 170, 182-84; and Lawson, *History of North Carolina*, 24.

33. Quotes from Catesby, *Natural History*, 24-26; and John Archdale, "A New Description of That Fertile and Pleasant Province of Carolina . . . ," in B.H. Carroll, *Historical Collections of South Carolina* (New York, 1836), 89-99.

34. Jennings, *Invasion of America*, 27; Waring, *History of Medicine*, 37-38; Bolzius, "Reliable Answer," 228. Darwin quote from Alfred W. Crosby, "Ecological Imperialism: The Overseas Migration of Western Europeans as a Biological Phenomenon," *Texas Quarterly* 21 (Spring 1978): 12; Thornton, Warren, and Miller, "Depopulation in the Southeast after 1492," 193.

Chapter 3

1. Bruce, *Economic History*, 1:122; Byrd, *London Diary*, 566; Lawson, *History of North Carolina*, 119, 121-23; A.W. Schorger, *The Wild Turkey: Its History and Domestication* (Nor-

man, Okla., 1966), 59-60; and idem, *The Passenger Pigeon: Its Natural History and Extinction* (Madison, Wis., 1955), 199-201.

2. Thomas A. Lund, "British Wildlife Law before the American Revolution," *Michigan Law Review* 74 (Nov. 1975): 49-74. Quote from *American Wildlife Law* (Berkeley, 1980), 8, by the same author. See also Robert M. Alison, "The Earliest Traces of a Conservation Conscience," *Natural History* 90 (1981): 72-77.

3. Lund, *American Wildlife Law*, 21-31.

4. Smith, *Generall Historie*, 147.

5. Walter P. Taylor, ed., *The Deer of North America ... Their History and Management* (Washington, D.C., 1956), 58; Schorger, *Wild Turkey*, 3-18, 142-43, 146-47; Lawson, *History of North Carolina*, 23.

6. Quoted in Arlie W. Schorger, *The Passenger Pigeon: Its Natural History and Extinction* (Madison, Wis., 1955), 5, 10-11; Lawson, *History of North Carolina*, 42.

7. Lawson, *History of North Carolina*, 120, 123; Byrd, *London Diary*, 546-47.

8. Taylor, *Deer of North America*, vii, 2, 18-19, 32, 57, 72-73, 138 (quotes from 57, 138); T.S. Palmer, *Private Game Preserves and Their Future in the United States*, Biological Survey Circular No. 72 (Washington, D.C., 1910), 2-11.

9. Quotes from Lawson, *History of North Carolina*, 146; and Byrd, *London Diary*, 566.

10. Thomas Cooper, *Statutes at Large ... of South Carolina* (Columbia, 1836-1841), 2:179 (hereafter cited as *South Carolina Laws*); William W. Hening, *The Statutes at Large: Being a Collection of All the Laws of Virginia* (Richmond, 1809), 4:446; 8:389-90 (hereafter cited as *Virginia Laws*); William Kilty, *The Laws of Maryland* (Annapolis, 1800), October 1728, chap. 7; 1758, chap. 11 (hereafter cited as *Maryland Laws*). On colonial laws, north and south, see also Yasuhide Kawashima and Ruth Tone, "Environmental Policy in Early America," *Journal of Forest History* 27 (Oct. 1983): 168-79.

11. Hening, *Virginia Laws*, 2:178, 215, 274-76, 282; 3:43, 141-42, 282-83; 4:89-90, 354; 6:152-53; 8:147-48, 388, 596. See also Richard G. Lillard, *The Great Forest* (New York, 1947), 75-76.

12. Kilty, *Maryland Laws*, 1758, chap. 11; 1790, chap. 8; Cooper, *South Carolina Laws*, 2:108; Craven, *White, Red, and Black*, 59-60.

13. Bruce, *Economic History*, 1:128, 371, 483-84; Lawson, *History of North Carolina*, 3; Beverly, *History*, 318; Lund, *American Wildlife Law*, 31.

14. Lawson, *History of North Carolina*, 8. See also Peter Matthiesen, *Wildlife in America* (New York, 1959), 42; Philip L. Barbour, "Further Notes on the Bison in Early Virginia," *Quarterly Bulletin of the Archeological Society of Virginia* 27 (1972): 100.

15. Clifford J. Hynning, *State Conservation of Resources* (Washington, D.C., 1939), 2, 23; Van Brocklin, "Conservation before 1901," 84.

16. Paul C. Phillips, *The Fur Trade* (Norman, Okla., 1961), 1:163-73; Murray G. Lawson, *Fur: A Study in English Mercantilism* (Toronto, 1943), 3, 53-54; Nathaniel C. Hale, *Pelts and Palisades* (Richmond, 1959), 61-62, 65. On some of the problems of the leather trade, see Hening, *Virginia Laws*, 1:488-89; 2:124-25, 185, 287, 482-83. Early South Carolina duties are recorded in Cooper, *South Carolina Laws*, 2:64.

17. J.R. Alden, *John Stuart and the Southern Colonial Frontier: A Study of Indian Relations, War, Trade, and Land Problems in the Southern Wilderness, 1754-1775* (New York, 1966), 17-18.

18. Byrd, *London Diary*, 572.

19. Catesby, *Natural History*, 22.

20. Lillard, *Great Forest*, 43-45, 61-63.

21. See Lawson, *History of North Carolina*, 35, 56. On extinctions, see tables in Paul A. Opler, "The Parade of Passing Species: A Survey of Extinctions in the U.S.," *Science Teacher* 44 (Jan. 1977).

22. Quote from Bolzius, "Reliable Answer," 228; Jenks Cameron, *The Development of*

Governmental Forest Control in the United States (Baltimore, 1928), 18, 19-20; Lillard, *Great Forest*, 67-68.

23. Lillard, *Great Forest*, 114-15.

24. Ronald L. Pollitt, "Wooden Walls: English Seapower and the World's Forests," *Journal of Forest History* (April 1971), 15.

25. Cf., e.g., Allen D. Candler, *The Colonial Records of the State of Georgia: Statutes Colonial and Revolutionary* (Atlanta, 1911), 18:73-74, 304 (hereafter cited as *Georgia Laws*); Hening, *Virginia Laws*, 1:332; Henry Potter, *Laws of the State of North Carolina* (Raleigh, 1821), 1:344 (hereafter cited as *North Carolina Laws*).

26. Lillard, *Great Forest*, 99-100; Jay P. Kinney, *Forest Legislation in America prior to March 4, 1789*, Agriculture Experiment Station Bulletin No. 370 (Ithaca, N.Y., 1916), 23, 26, 370; Kilty, *Maryland Laws*, Nov. 1792 and Jan. 15, 1799; Samuel Trask Dana, *Forest and Range Policy: Its Development in the United States* (New York, 1956), 6-7; Potter, *North Carolina Laws*, 1:347-48, 434; Candler, *Georgia Laws*, 18:556-57.

27. Lillard, *Great Forest*, 67-68; Merrens, *Colonial North Carolina*, 85-86, 93, 107. A similar judgement on a part of the Tidewater is reached by Earle, *Evolution of a Tidewater Settlement*, 34.

28. Bolzius, "Reliable Answer," 229, 231-32, 238, 241, 258.

29. Hening, *Virginia Laws*, 1:199.

30. Ibid., 3:180.

31. Ibid., 3:462-63; 5:60-63: 7:412-13.

32. Kilty, *Maryland Laws*, May 1730, chap. 17; Nov. 1773, chap. 24; 1785, chap. 29; 1789, chap. 5; 1795, chap. 49.

33. Potter, *North Carolina Laws*, 1:128-29, 169-70, 239, 263, 392; Cooper, *South Carolina Laws*, 4:310-12, 411, 719.

34. Hening, *Virginia Laws*, 2:30; Cooper, *South Carolina Laws*, 3:269; Kilty, *Maryland Laws*, 1771, chap. 6; 1777, chap. 17; 1784, chap. 83; 1791, chap. 84; 1798, chap. 71. The Georgia law of 1773 was more concerned with cattle than deer, however. See Candler, *Georgia Laws*, 19:288-90.

35. Byrd, *London Diary*, 547.

36. Jack P. Green, ed., *The Diary of Colonel Landon Carter of Sabine Hall, 1752-1778* (Charlottesville, Va., 1965), 1:8-9, 256-57. On crop diversification in the Tidewater as tobacco prices fell, see Paul G.E. Clemens, "The Operation of an 18th Century Chesapeake Tobacco Plantation," *Agricultural History* 49 (July 1975): 517-31.

37. Henry Clepper et al., *Origins of American Conservation* (New York, 1966), 92. See also Van Brocklin, "Conservation before 1901," 119-24; Petulla, *American Environmental History*, 56-57.

38. Cf. Craven, *Soil Exhaustion*, 190-91.

39. Jefferson, *Notes on the State of Virginia* (Boston, 1801), 65.

40. James A. Tobey, *Public Health Law* (Baltimore, 1926), 10; Berkeley and Berkeley, *Dr. Alexander Garden*, 137.

41. See Roderick Nash, *Wilderness and the American Mind* (New Haven, Conn., 1967, 1973), 51-53; Sharon T. Pettie, "Preserving the Great Dismal Swamp,170 *Journal of Forest History* 20 (Jan. 1976): 28-30.

42. Quote from Edmund Berkeley and Dorothy S. Berkeley, *John Clayton, Pioneer of American Botany* (Chapel Hill, 1963), 51. On the English gardening mania, see Richard Beale Davis, *Intellectual Life in the Colonial South, 1585-1763* (Knoxville, Tenn., 1978), 2:812.

43. See George Frick's introduction to Catesby, *Natural History*, ix-xv; George Frederick Frick and Raymond Phineas Stearns, *Mark Catesby, the Colonial Audubon* (Urbana, Ill., 1961), 64. The Stono quote is in Catesby, *Natural History*, 17.

44. Lawson, *History of North Carolina*, 134-35; Henry Savage, Jr., *Discovering America*,

1700-1875 (New York, 1979), 19-24; Joseph Kastner, *A Species of Eternity* (New York, 1977), 15-16.

45. Kastner, *Species of Eternity*, 58, 66-67; Berkeley and Berkeley, *John Clayton*, 33.

46. Quotes from Berkeley and Berkeley, *Dr. Alexander Garden*, 123, 125; and G. Edmund Gifford, Jr., "The Charleston Physician Naturalists," *Bulletin of the History of Medicine* (Winter 1975), 559. See also Duffy, *Epidemics in Colonial America*, 9, 36.

47. Quoted in Kastner, *Species of Eternity*, 97.

48. Savage, *Discovering America*, 69; John Livingston Lowes, *The Road to Xanadu: A Study in the Ways of the Imagination* (London, 1927, 1951); 364-72; Coleridge quote from Norman Fruman, *Coleridge, the Damaged Archangel* (New York, 1971), 194, 525n. For Bartram's early observation of "waters breaking out of the earth," see his "Travels in Georgia and Florida, 1773-74: A Report to Dr. John Fothergill," *Transactions of the American Philosophical Society* 33 (Nov. 1943): 156.

Chapter 4

1. Kulikoff, "Tobacco and Slaves," 47-48; Gray, *History of Agriculture*, 2:121-22, 125-26.

2. Gray, *History of Agriculture*, 2:686.

3. John Filson, *The Discovery, Settlement and Present State of Kentucke . . .* (Wilmington, 1784), 12.

4. Taylor, *Deer of North America*, 20; Gray, *History of Agriculture*, 2:861.

5. "A Memorandum of Moses Austin's Journey . . . 1796-1797," *American Historical Review* 5 (Apr. 1900): 27; Ray Allen Billington, *Westward Expansion: A History of the American Frontier* (New York, 1960), 160-73.

6. John Duffy, *The Rudolph Matas History of Medicine in Louisiana* (Baton Rouge, 1958), 1:10 (hereafter cited as *Medicine in Louisiana*); M. Penicaut, "Annals of Louisiana from 1698 to 1722," in B.F. French, *Historical Collections of Louisiana and Florida* (New York, 1869), 138-39. Quote from Gray, *History of Agriculture*, 1:61.

7. Quote from David G. Wilson, "Range and Forest Resources," in Clepper *Origins*, 36; Duffy, *Medicine in Louisiana*, 1:127; Gray, *History of Agriculture*, 1:63, 65, 78.

8. Brown, *Historical Geography*, 40.

9. P. de Charlevoix, *Journal of a Voyage to North-America . . .* (London, 1761), 2:282; Dumont de Montigny, "History of Louisiana," in *Historical Collections of Louisiana* (New York, 1853), 23-24.

10. Gray, *History of Agriculture*, 1:76-77, 79; Joseph C. Tregle, ed., *The History of Louisiana, Translated from the French of M. Le Page du Pratz* (Baton Rouge, 1975), 254.

11. Duffy, *Medicine in Louisiana*, 1:128-30, 132, 206-7, 347.

12. Charlevoix, *Journal*, 2:284; John R. Swanton, *Indian Tribes of the Lower Mississippi Valley and Adjacent Coast of the Gulf of Mexico* (Washington, D.C., 1911), 101; Thomas Nuttall, *Journal of Travels in the Arkansas Territory during the Year 1819 . . .* (Philadelphia, 1821), 58; Duffy, *Medicine in Louisiana*, 1:22, 38.

13. Duffy, *Medicine in Louisiana*, 1:232-33; James Penick, Jr., *The New Madrid Earthquakes of 1811-12* (Columbia, Mo., 1976), 48-49, 119; Wayne Viitanen, "The Winter the Mississippi Ran Backwards: Early Kentuckians Report the New Madrid, Missouri, Earthquake of 1811-12," *Register of the Kentucky Historical Society* 71 (Jan. 1973): 51-68; Arch C. Johnston, "A Major Earthquake Zone on the Mississippi," *Scientific American* 246 (Apr. 1982): 60-83.

14. Quote from Gray, *History of Agriculture*, 2:289-90.

15. Ibid., 2:888-89.

16. Ibid., 2:687-88, 691.

17. Quote in William C. Bagley, Jr., *Soil Exhaustion and the Civil War* (New York, 1942), 32-33. See also Edward C. Papenfuse, "Planter Behavior and Economic Opportunity in a Staple Economy," *Agricultural History* 46 (Apr. 1973): 299.

18. Billington, *Westward Expansion*, 31; U.S. Bureau of the Census, *The Statistical History of the U.S. from Colonial Times to the Present* (New York, 1976), 1168; quote in James L. Patton, "Letters from North Carolina Emigrants in the Old Northwest, 1830-1834," *Mississippi Valley Historical Review* 47 (Sept. 1960): 266.

19. Malcolm J. Rohrbaugh, *The Transappalachian Frontier* (New York 1978), 279; Nuttall, *Travels*, 78, 92.

20. L.P. Hebert, *Culture of Sugarcane for Sugar Production in Louisiana*, USDA Handbook No. 262 (Washington, D.C., 1964), 19; J.F.C. Hagens, *Fertilizing Sugar Cane* (New York, n.d.).

21. Quote from Mary C. Gillett, *The Army Medical Department, 1818-1865* (Washington, D.C., 1987), 71.

22. Gray, *History of Agriculture*, 2:908-9.

23. Stanley W. Trimble, *Man-Induced Soil Erosion on the Southern Piedmont, 1700-1970* (Washington, D.C., 1974), especially p. 16; Arthur C. Hall, *The Story of Soil Conservation in the South Carolina Piedmont, 1800-1860*, USDA Miscellaneous Publication 407 (1940), 1-3.

24. Cf. Gray, *History of Agriculture*, 2:882.

25. Julius Rubin, "The Limits of Agricultural Progress in the Nineteenth-Century South," *Agricultural History* 49 (Apr. 1975): 362-74. On the appetites of corn, see Sprague, *Corn and Corn Improvement*, 632-35.

26. Drew Gilpin Faust, "The Rhetoric and Ritual of Agriculture in Antebellum South Carolina," *Journal of Southern History* 45 (Nov. 1979): 541-68; David W. Francis, "Antebellum Agricultural Reform in *DeBow's Review*," *Louisiana History* 14 (Spring 1973): 165-78; G. Melvin Herndon, "Agricultural Reform in Antebellum Virginia: William Galt, Jr.: A Case Study," *Agricultural History* 52 (July 1978): 394-406.

27. V.R. Cardozier, *Growing Cotton* (New York, 1957), 101ff; R.P. Bartholomew, *Soil Improvement Practices Afecting Yields of Cotton*, Bulletin 513, Arkansas Agricultural Experiment Station (Fayetteville, 1951), 134; Richard C. Sheridan, "Chemical Fertilizers in Southern Agriculture," *Agricultural History* 53 (Jan. 1979): 308-18. On Ruffin, see Craven, *Soil Exhaustion*, 141; and J. Carlyle Sitterson, ed., *Essay on Calcareous Manures, by Edmund Ruffin* (Cambridge, 1961).

28. Quote from Richard K. Vedder ad David C. Stockdale, "The Profitability of Slavery Revisited: A Different Approach," See also William J. Cooper, Jr., "The Cotton Crisis in the Antebellum South: Another Look," both in *Agricultural History*, 49 (Apr. 1975): 381-404; Eugene D. Genovese, *The Political Economy of Slavery* (New York, 1967), 85-99, 117-18; Robert E. Gallman, "Self-Sufficiency in the Cotton Economy of the Antebellum South," *Agricultural History* 44 (Jan. 1970): 5-23; Mark D. Schmitz, "Farm Interdependence in the Antebellum Sugar Sector," *Agricultural History* 52 (Jan. 1978): 92-103.

29. See Stanley L. Engerman, "A Reconsideration of Southern Economic Growth, 1770-1860," and Harold D. Woodman, "New Perspectives on Southern Economic Development: A Comment," both in *Agricultural History* 49 (Apr. 1975): 343-61, 374-80. Quote from 380.

30. Cf. Robert W. Fogel and Stanley L. Engerman, *Time on the Cross: The Economics of American Negro Slavery* (Boston, 1974), 1:167-68; U.S. Congress, Senate, *Senate Executive Document* 407, 28 Cong., 1 ess. (1844) 35-54.

Chapter 5

1. Quotes from Henry D. Shapiro and Zane L. Miller, eds., *Physician to the West: Selected Writings of Daniel Drake on Science and Society* (Lexington, Ky., 1970), 355, 379; Daniel Drake, *A Systematic Treatise, Historical, Etiological, and Practical of the Principal Diseases of the Interior Valley of North America* (1850-1854; rpt. New York, 1971), 2:2.

2. Nuttall, *Travels*, 213-14.

3. Ibid., 35, 230; Shapiro and Miller, *Physician to the West*, 359.

4. Dale C. Smith, "Quinine and Fever: The Development of the Effective Dosage," *Journal of the History of Medicine and the Allied Sciences* 31 (July 1976): 343-67; Boyd, *Malariology*, 13; William R. Horsfall, *Mosquitoes: Their Bionomics and Relation to Disease* (New York, 1972), 149-50.

5. Duffy, *Medicine in Louisiana*, 1:484. Gillett, "Army Medical Department," chapters 2 and 3. Fort Gibson, in what is now Oklahoma, later succeeded to this unenviable distinction.

6. Quote from Savitt, *Medicine and Slavery*, 59.

7. Josiah C. Nott, "Life Insurance at the South," *DeBow's Commerical Review of the South and West* 3 (May 1847): 366. On the impact of epidemic disease, see also: J.C. Nott, "Sketch of the Epidemic of Yellow Fever of 1847, in Mobile," *Charleston Medical Journal and Review* 3 (Jan. 1848): 1-20; Gordon Gillson, "Louisiana: Pioneer in Public Health," *Louisiana History* 4 (1963): 207-32; Douglas F. Stickle, "Disease and Class in Baltimore: The Yellow Fever Epidemic of 1800," *Maryland Historical Magazine* 74 (Sept. 1979): 282-99; George F. Pearce, "Torment of Pestilence: Yellow Fever Epidemics in Pensacola," *Florida Historical Quarterly* 56 (Apr. 1978): 448-72.

8. On cholera in the South, see Nancy D. Baird, "Asiatic Cholera's First Visit to Kentucky: A Study in Panic and Fear," *Filson Club History Quarterly* 48 (July 1974): 228-40.

9. See John H. Ellis, "Business and Public Health in the Urban South During the Nineteenth Century: New Orleans, Memphis, and Atlanta," *Bulletin of the History of Medicine* 44 (1970): 197-212. See also Duffy, *Medicine in Louisiana*, 2: 162; David R. Goldfield, "The Business of Health Planning: Disease Prevention in the Old South," *Journal of Southern History* 42 (Nov. 1976): 557-70; JoAnn Carrigan, "Impact of Epidemic Yellow Fever on Life in Louisiana," *Louisiana History* 4 (Winter 1963): 7-8.

10. Quote from Nott, "Life Insurance," 66.

11. First Bromme quote from Louis E. Brister, "The Image of Arkansas in the Early German Immigrant Guidebook: Notes on Immigration," *Arkansas Historical Quarterly* 36 (Winter 1977): 342-43; and Bromme's *Hand- und Reisebuch für Auswanderer und Reisende nach Nord-Mittel- und Süd-Amerika* (Bamberg, 1853), 248, 251.

12. Rev. Mr. Price [not otherwise identified], "The Mississippi Swamp," *DeBow's Commercial Review of the South and West* 7 (July 1849): 55.

13. Clepper, *Origins*, 42; G. Melvin Herndon, "The Significance of the Forest to the Tobacco Plantation Economy in Antebellum Virginia," *Plantation Society in the Americas* (Oct. 1981), 430-39.

14. Percival Perry, "The Naval-Stores Industry in the Old South, 1790-1860," *Journal of Southern History* 34 (Nov. 1968): 509-26. The exact process of extracting turpentine as practiced in contemporary South Carolina is given by Edwin Heriot, "Manufacture of Turpentine in the South," *DeBow's Commercial Review of the South and West* 8 (May 1850): 452-54.

15. Clepper, *Origins*, 41; Van Brocklin, Conservation Before 1901," 11-17, and sources cited therein.

16. Carl H. Clendening, "Early Days in the Southern Appalachians," *Southern Lumberman* 144 (1931): 101-05.

17. Quoted in John H. Goff, "The Great Pine Barrens," *Emory University Quarterly*, 5 (March 1949): 23.

18. Thomas C. Nelson, "The Original Forests of the Georgia Piedmont," *Ecology* 38 (July 1957): 390-97.

19. Quote from Nuttall, *Travels*, 58.

20. John A. Eisterhold, "Lumber and Trade in the Lower Mississippi Valley and New Orleans, 1800-1860," *Louisiana History* 13 (Winter 1972): 71-72.

21. Gerald L. Collier, "The Evolving East Texas Woodland" (PhD. diss. University of Nebraska, 1964); Corliss C. Curry, "Early Timber Operations in Southeast Arkansas," *Arkansas Historical Quarterly* 19 (Summer 1960): 111-18.

22. Henry Toulmin, *Digest of the Laws of the State of Alabama* (Cahawba, 1823), March 1, 1803 (hereafter cited as *Alabama Laws*); R.H. Clark et al., *The Code of the State of Georgia* (Atlanta, 1867), 284-85, 321-22, 652, 891; Arthur Foster, *A Digest of the Laws of the State of Georgia* (Philadelphia, 1831), 129; E.H. English, *A Digest of the Statutes of Arkansas* (Little Rock, 1848), 1002-3; William A. Hotchkiss, *Codification of the Statute Law of Georgia* (Savannah, 1845), 763-65; Preston S. Loughborough, *A Digest of the Statute Laws of Kentucky* (Frankfort, 1842), 133-34; Oliver H. Prince, *Digest of the Laws of the State of Georgia* (Athens, Ga., 1837), 692; Allen H. Bush, comp., *A Digest of the Statute Law of Florida* (Tallahassee, 1872), 98-99, 335-37; William L. Martin, *The Code of Alabama* (Atlanta, 1897), 1:211; Edward W. Gantt, *A Digest of the Statutes of Arkansas* (Little Rock, 1874), 346; James F. McClellan, *Digest of the Laws of the State of Florida* (Tallahassee, 1881): 984-85 (hereafter cited as *Florida Laws*). Firehunting (not in the sense of firing the woods, but of night hunting with a fire pan to spot the deer by the reflection of its eyes) is described in detail in Henry M. Peck, "Deer and Deer Hunting in Louisiana," *DeBow's Commerical Review of the South and West* 5 (March 1848): 227-29. For an excellent general account of hunting in this period, see Sam Bowers Hilliard, *Hog Meat and Hoecake: Food Supply in the Old South* (Carbondale, Ill., 1972), 70-83, on which the account here given is chiefly based.

23. William Elliott, *Carolina Sports by Land and Water* (New York, 1967): 150-51.

24. Ibid., 168.

25. Sydnor, *Southern Sectionalism*, 273-74 n.

26. Forrest G. Hill, *Roads, Rails and Waterways: The Army Engineers in Early Transporation* (Norman, Okla., 1957), 207-8.

27. On *Gibbons v. Ogden* and the Commerce Clause, see Felix Frankfurter, *The Commerce Clause under Marshall, Taney and Waite* (Chicago, 1964).

28. Gray, *History of Agriculture*, 2:902-3. See also Thomas E. Jeffrey, "Internal Improvements and Political Parties in Antebellum North Carolina, 1836-1860." *North Carolina Historical Review* 55 (Apr. 1978): 111-56.

29. Andrew A. Humphreys and Henry L. Abbot, *Report upon the Physics and Hydraulics of the Mississippi River . . .* (Washington, D.C., 1876), 428-45.

30. Quotes from Nuttall, *Travels*, 48, 61; and Hynning, *State Conservation*, 57.

31. John Ise, *United States Forest Policy* (New Haven, 1920), 38; U.S. Congress, Senate, Senate Executive Document 55, 50 Cong., 1 sess. (1888), 8-9; *Congressional Globe*, 30 Cong., 2 sess. (1848), 46, 120; Rachel Edna Norgress, "The History of the Cypress Lumber Industry in Louisiana," *Louisiana Historical Quarterly* 30 (June 1947): 979-1059; Van Brocklin, "Conservation Before 1901," 242; Paul W. Gates, "Federal Land Policies in the Southern Public Land States," *Agricultural History* 53 (Jan. 1979): 206. The function of the swamps in retaining water is clearly set forth by a Louisiana doctor, H.D. Peck, in "The Levee System of Louisiana," *DeBow's Review* 8 (Feb. 1850): 101-4.

32. Charles S. Sydnor, "State Geological Surveys in the Old South," in *American Studies in Honor of William Kenneth Boyd* (Freeport, N.Y., 1968), 86-108; Hynning, *State Conservation*, 41; *The State Geological Surveys of the United States*, USGS Survey Bulletin No. 465 (Washington, D.C. 1911).

33. Hynning, *State Conservation*, 35, 38; William C. Dawson, *Compilation of the Laws of the State of Georgia* (Milledgeville, Ga., 1831), 286.

34. Charles Ellet, *The Mississippi and Ohio Rivers: Containing Plans for the Protection of the Delta* (Philadelphia, 1853).

35. See Theodore Dwight Bozeman, "Joseph LeConte: Organic Science and a 'Sociology for the South,'" *Journal of Southern History* 39 (Nov. 1973): 565-82; James Dennis

Guillory, "The Pro-Slavery Arguments of Dr. Samuel A. Cartwright," *Louisiana History* 9 (Summer 1968): 209-28; idem, "Diversity of the Human Race," *DeBow's Review* 10 (Feb. 1851): 114; Carl N. Degler, *Place over Time: The Continuity of Southern Distinctiveness* (Baton Rouge, 1977), 90; Ronald L. Numbers and Janet S. Numbers, "Science in the Old South: A Reappraisal," *Journal of Southern History* 48 (May 1982): 163-84.

36. Among useful works in this area are: Francis H. Herrick, *Audubon the Naturalist: A History of His Life and Time* (New York, 1938); Thomas Nuttall, *The American Ornithology* (Philadelphia, 1808-1814); Robert H. Welker, *Birds and Men: American Birds in Science, Art, Literature and Conservation* (Cambridge, Mass., 1953); John Abbot, "Autobiography," ed. C.L. Remington, in *Lepidopterist News* 2 (Mar. 1948); Jeanette E. Graustein, *Thomas Nuttall, Naturalist: Explorations in America, 1808-1841* (Cambridge, Mass., 1967). The Audubon quotes are from Robin W. Doughty, *Feather Fashions and Bird Preservation* (Berkeley, 1975), 38. See also Roger Tory Peterson, introduction, *The Art of Audubon: The Complete Birds and Mammals* (New York, 1980), x-xi. There is a valuable appreciation of Audubon the artist in Welker, *Birds and Men*, 85-89. Curious and perceptive is Eudora Welty's "A Still Moment," in her *Collected Stories* (New York, n.d.), 189-99.

Chapter 6

1. John A. Garraty, *The New Commonwealth, 1877-1890* (New York, 1968), 84-85.
2. E.C. Faust, "The History of Malaria in the United States," *American Scientist* 39 (1951): 122; Shryock, *Development of Modern Medicine*, 240-41.
3. Paragraphs based on: Richard H. Shryock, "The Origins of the American Public Health Movement," *Annals of Medical History* n.s. 1 (Nov. 1929): 661, especially notes 44, 45. On racist medical thought, see John S. Halle, Jr., "The Physician versus the Negro: Medical and Anthropological Concepts of Race in the Late Nineteenth Century," *Bulletin of the History of Medicine* 44 (1970): 154-67; and idem, "Race, Mortality and Life Insurance: Negro Vital Statistics in the Late Nineteenth Century," *Journal of the History of Medicine and Allied Sciences* 25 (July 1970): 247-61; Duffy, *Medicine in Louisiana*, 2:449. See also Marshall Scott Legan, "Disease and the Freedmen in Mississippi during Reconstruction," *Journal of the History of Medicine and Allied Sciences* 28 (July 1973): 261-63. Quote on drug addiction from David T. Courtwright, *Dark Paradise: Opiate Addiction in America before 1940* (Cambridge, Mass., and London, 1982), 38-39.
4. Quotes in Thomas H. Baker, "Yellowjack: The Yellow Fever Epidemic of 1878 in Memphis, Tennessee," *Bulletin of the History of Medicine* 42 (May-June 1968): 264. See also Gerald M. Capers, Jr., "Yellow Fever in Memphis in the 1870's," *Mississippi Valley Historical Review* 24 (1938): 483-502; Dennis East, "Health and Wealth: Goals of the New Orleans Public Health Movement, 1879-84," *Louisiana History* 9 (Summer 1968): 250, 256. On the diffusion of yellow fever from Memphis and New Orleans to the hinterland, see, e.g., Marshall Scott Legan, "Mississippi and the Yellow Fever Epidemic of 1878-1879," *Journal of Mississippi History* 33 (Aug. 1971): 199-218.
5. Gavin Wright and Howard Kunreuther, "Cotton, Corn and Risk in the Nineteenth Century," *Journal of Economic History* 35 (Sept. 1975): 527; Gilbert C. Fite, "Southern Agriculture since the Civil War. An Overview," *Agricultural History* 53 (Jan. 1979): 9.
6. P.M. Ashburn, *A History of the Medical Department of the United States Army* (Boston, 1929), 168-69: Scheffel H. Wright, "Medicine in the Florida Camps during the Spanish American War—Great Controversies," *Journal of the Florida Medical Association* 62 (Aug. 1975): 21-23.
7. Harrison, *Mosquitoes, Malaria and Man*, 96-99.
8. Quotes from Fite, "Southern Agriculture since the Civil War," 7-8. See also Roger Ransom and Richard Sutch, "The 'Lock-In' Mechanism and Overproduction of Cotton

in the Postbellum South," *Agricultural History* 44 (Apr. 1975): 405-25; C. Vann Woodward, *Origins of the New South, 1877-1913* (Baton Rouge, 1951, 1971), especially 175-205; Robert L. Brandfon, *The Cotton Kingdom of the New South* (Cambridge, Mass., 1967), 3, 15; Merle Prunty, Jr., "The Renaissance of the Southern Plantation," *Geographical Review* 45 (Oct. 1955): 460-63; E.A. Boeger and E.A. Goldenweiser, *A Study of the Tenant Systems of Farming in the Yazoo-Mississippi Delta*, USDA Bulletin No. 337 (Washington, D.C., 1916), 2; Stephen DeCanio, "Productivity and Income Distribution in the Post-Bellum South," *Journal of Economic History* 34 (June 1974): 422-26; Kenneth S. Greenberg, "The Civil War and the Redistribution of Land, Adams County, Mississippi, 1860-1870," *Agricultural History* 52 (Apr. 1978): 292-307. On banks, see John A. James, "Financial Underdevelopment in the Postbellum South," *Journal of Interdisciplinary History* 11 (Winter 1981): 443-54.

9. Bruce, *Economic History*, 1:153-54.

10. See Lester D. Stephens, "Farish Furman's Formula: Scientific Farming and the 'New South,'" *Agricultural History* 50 (Apr. 1976): 377-90; Kenneth Coleman, ed., "How to Run a Middle Georgia Cotton Plantation in 1885: A Document," *Agricultural History* 42 (Jan. 1968): 55-60: Fred A. Shannon, *The Farmer's Last Frontier: Agriculture, 1860-1897* (New York, 1945), especially 92-124: Willard Range, "P.J. Berckmans: Georgia Horticulturist," *Georgia Review* 6 (Summer 1952): 222; Eugene H. Lerner, "Southern Output and Agricultural Income, 1860-1880," *Agricultural History* 33 (July 1959): 117-25; Fite, "Southern Agriculture since the Civil War," 11.

11. See C. Golden and F. Lewis, "The Economic Costs of the American Civil War: Estimates and Implications," *Journal of Economic History* 35 (June 1975): 299-326.

12. Donald J. Millet, "Some Aspects of Agricultural Retardation in Southwest Louisiana, 1865-1900," *Louisiana History* 11 (Winter 1970): 37-62; J. Carlyle Sitterson, *Sugar Country: The Cane Sugar Industry in the South, 1753-1950* (Lexington, Ky., 1953), 252-68.

13. Millet, "Some Aspects," 167-242; W. Rodney Cline, "Seaman Asahel Knapp, 1833-1911," *Louisiana History* 11 (Fall 1970): 333-40. On the collapse of North Carolina's rice industry in the face of Louisiana competition and a series of devastating hurricanes, see Clifton, "Golden Grains," 385-86.

14. Harrison, *Mosquitoes, Malaria and Man*, 58-60; Jane M. Porter, "Experiment Stations in the South, 1877-1940," *Agricultural History* 53 (Jan. 1979): 91-93. On the diseases that long plagued southern cattle, see Tamara Miner Haygood, "Cows, Ticks, and Disease: A Medical Interpretation of the Southern Cattle Industry," *Journal of Southern History* 52 (Nov. 1986): 551-64.

15. Brandfon, *New South*, 85.

16. John Douglas Helms, "Just Lookin' for a Home: The Cotton Boll Weevil and the South" (Ph.D. diss., Florida State University, 1977); W.D. Hunter and W.E. Hinds, *The Mexican Cotton Boll Weevil*, USDA, Bureau of Entomology, Bulletin No. 51 (Washington, D.C., 1905); Thomas R. Dunlap, *DDT: Scientists, Citizens, and Public Policy* (Princeton, 1981), 25-29.

17. Quote from Paul Wallace Gates, "Federal Land Policy in the South, 1866-1888," *Journal of Southern History* 6 (Aug. 1940): 316; Claude F. Oubre, "Forty Acres and a Mule: Louisiana and the Southern Homestead Act," *Louisiana History* 17 (Spring 1976): 143-58; *U.S. Statutes at Large*, 14 (1866-67): 66; 19:73.

18. *U.S. Statutes at Large*, 25 (1887-89): 854; Gates, "Federal Land Policy," 311-30; Warren Hoffnagle, "The Southern Homestead Act: Its Origin and Operation," *Historian* 32 (Aug. 1970): 612-29. On the national transfer of forest resources to private hands, see U.S. Congress, *Senate Document 818*, 61 Cong., 3 sess. (1911), 3. Quote from Paul Wallace Gates, "Federal Land Policies in the Southern United States," *Agricultural History* 53 (Jan. 1979): 219-20.

19. Thomas D. Clark, "The Impact of the Timber Industry on the South," *Mississippi*

Quarterly 25 (Spring 1972): 141-64, especially 143-60. For a sensitive personal account of "cut and get out" logging in a southern mill town, see Otis Dunbar Richardson, "Fullerton, Louisiana: An American Monument," *Journal of Forest History* 27 (Oct. 1983): 192-201. The forestland devastated at Fullerton later became part of Kisatchie National Forest. The New South credo of development at all costs had its effect on the wetlands as well. See C.T. Trowell and P.L. Izlar, "Jackson's Folly: The Suwanee Canal Company in the Okefenokee Swamp," *Journal of Forest History* 27 (Oct. 1984): 187-95.

20. Quote from J.M. Stauffer, "The Timber Resources of 'The Southwest Alabama Forest Empire,'" *Journal of the Alabama Academy of Science* 30 (Jan. 1959): 52-67.

21. Quote from Goff, "Pine Barrens," 30.

22. On this topic, see also Collier, "East Texas Woodland"; Corliss C. Curry, "Early Timber Operations in Southeast Arkansas," *Arkansas Historical Quarterly* 19 (Summer 1960): 111-18; Thomas D. Clark, "Early Lumbering Activities in Kentucky," *Northern Logger* 13 (Mar. 1965): 14-15, 42-43; Carl H. Clendening, "Early Days in the Southern Appalachians," *Southern Lumberman* 144 (Dec. 15, 1931): 101-5; Donald Joseph Millet, Sr., "The Economic Development of Southwest Louisiana, 1865-1900" (Ph.D. diss., Louisiana State University, 1964), 260. On Alabama's iron industry, see Gates, "Land Policy in the South," 217; and the following articles in *Alabama Review*: John B. Ryan, "Willard Warner and the Rise and Fall of the Iron Industry in Tecumseh, Alabama," 24 (Oct. 1976): 261-79; Hugh C. Bailey, "Ethel Armes and *The Story of Coal and Iron in Alabama*," 22 (July 1969): 188-99; Robert H. McKenzie, "Reconstruction of the Alabama Iron Industry, 1865-1880," 25 (July 1972): 178-91. The story is carried forward in George B. Tindall, *The Emergence of the New South, 1913-45* (Baton Rouge, 1967), 57, 81.

23. Schorger, *Passenger Pigeon*, 214-30. Interesting also is David and Jim Kimball, *The Market Hunter* (Minneapolis, 1969), a casual but perceptive essay. On Texas, see Robin W. Doughty, *Wildlife and Man in Texas: Environmental Change and Conservation* (College Station, Tex., 1983).

24. Clepper, *Origins*, 24; Mikko Saikku, "The Extinction of the Carolina Parakeet," *Environmental History Review* 14 (Fall 1990): 1-20.

25. Edward I. Bullock et al., *The General Statutes of the Commonwealth of Kentucky* (Frankfort, 1873), 365; John L. Hopkins et al., *The Code of the State of Georgia* (Atlanta, 1896), 1:463, 491; 3:160; William F. Kirby, *A Digest of the Statutes of Arkansas* (Austin, 1904), 824-29; John Prentiss Poe, *The Maryland Code: Public General Laws* (Baltimore, 1888), 2:1304-5; William A. Seay and John S. Young, *The Revised Statutes of the State of Louisiana* (Baton Rouge, 1886), 236-37 (hereafter cited as *Louisiana Laws*); McClelland, *Florida Laws*, 430; John D. Carroll, *Kentucky Laws: Civil and Criminal Codes of Practice* (Louisville, 1889), 749-51. See also Theodore W. Cart, "The Struggle for Wildlife Protection in the United States, 1870-1900" (Ph.D. diss., University of North Carolina, 1971), 57. The Mencken quote is from Alistair Cooke, ed., *The Vintage Mencken* (New York, 1955), 5.

26. Seay and Young, *Louisiana Laws*, 238-40; Noel Simon and Paul Géroudet, *Last Survivors: The Natural History of Animals in Danger of Extinction* (New York, 1970), 63.

27. John F. Reiger, *American Sportsmen and the Origins of Conservation* (New York, 1975), 31. See also Edward H. Graham, *The Land and Wildlife* (New York, 1947), 32; Edmund Morris, *The Rise of Theodore Roosevelt* (New York, 1979), 383-85; James B. Trefethen, *Crusade for Wildlife: Highlights in Conservation Progress* (New York, 1961), 15-20.

28. Clepper, *Origins*, 25; Cart, "Struggle for Wildlife," 31; Scott quote from Robin W. Doughty, *Feather Fashions and Bird Preservation: A Study in Nature Protection* (Berkeley, 1975), 81-82.

29. McClelland, *Florida Laws*, 429-30.

30. Clepper, *Origins*, 42-44.

31. Hynning, *State Conservation*, 17; *Annual Report of the Commissioner of the General Land Office for the Year 1884* (Washington, D.C., 1885), 19; *Annual Report of the Secretary of the*

Interior for the Year 1885 (Washington, D.C., 1886), 1:45-46. 236; James D. Richardson, *A Compilation of the Messages and Papers of the Presidents, 1789-1897* (Washington, D.C., 1908), 8:369, 521.

32. Herbert A. Smith, "The Early Forestry Movement in the United States," *Agricultural History* 12 (Oct. 1938): 332-33.

33. Harold T. Pinkett, *Gifford Pinchot, Private and Public Forester* (Urbana, Ill., 1970), especially 23-24.

34. Van Brocklin, "Conservation before 1901," 65, 78. This argument stands on uncertain ground in every sense; the action of forests on runoff is extremely complex.

35. U.S. Congress, Sen. Doc. 818, 61 Cong., 3 sess. (1911), 3.

36. *Report of the Chief of Engineers to the Secretary of War, 1869* (Washington, D.C., 1869), 327-52; ibid., *1875*, 557.

37. Newspaper clipping with printed letter, Gibson to Lourey [1881?], Folder 10, Benjamin M. Harrod Collection, Howard-Tilton Memorial Library, Tulane University, New Orleans, La.

38. U.S. Congress, Senate, *Miscellaneous Document 178*, 53 Cong., 2 sess. (1894), 79; *Congressional Record*, 45 Cong., 3 sess. (1879), 262-63; 49 Cong., 2 sess. (1887), 706.

39. *Congressional Record*, 46 Cong., 1 sess. (1879), 2283-84; *U.S. Statutes at Large*, 21 (1879-81): 37. On Garfield's intervention, see Gibson to his law partner (probably his brother, McKinley Gibson) in undated clipping, item no. 253, Folder 10, Harrod Collection.

40. See S.W. Ferguson to Taylor, Jan. 27 and 30, 1886, with enclosures, in Robert S. Taylor Collection, Indiana State Library, Indianapolis.

41. Quotes from New Orleans *Daily Picayune*, Mar. 24, 1884; and James B. Eads, *Physics and Hydraulics of the Mississippi River: Report of the U.S. Levee Commission Reviewed* (New Orleans, 1876), 31-32

42. See *Proceedings of the Mississippi River Commission*, Oct. 1, 1890, 1, 6, 16.

43. Mississippi River Commission (MRC), "Flood slopes for Mississippi River at maximum stages at Red River Landing, La., 1872 to 1912," National Archives, Record Group 77, Ser. 522.

44. The process can be followed in the *Annual Reports of the Chief of Engineers, 1892*, 4:2901-2; *1893*, 5:3545, 3557-58. See also *Proceedings of the MRC*, Aug. 2, 1892, 2; Jan. 11, 1896, 297. An interesting appreciation of Gibson's work is U.S. Congress, Senate, *Miscellaneous Document 178*, 52 Cong., 2 sess. (1894).

Chapter 7

1. Allison Davis et al., *Deep South: A Social Anthropological Study of Caste and Class* (Chicago, 1941), 158.

2. E. Dwight Sanderson, *A Statistical Study of the Decrease in the Texas Cotton Crop Due to the Mexican Cotton Boll Weevil and the Cotton Acreage of Texas, 1899-1904 Inclusive* (Austin, 1905).

3. Theodore Salutos, *Farmer Movements in the South, 1865-1933* (Berkeley, 1960), 227.

4. Quote from Clifton Paisley, "Madison County's Sea Island Cotton Industry, 1870-1916," *Florida Historical Quarterly* 54 (Jan. 1976): 303.

5. Quote from Rupert Vance, *Human Factors in the Cotton Culture: A Study in the Social Geography of the American South* (Chapel Hill, 1929), 25-26.

6. Arthur F. Raper, *Tenants of the Almighty* (New York, 1943), 158.

7. Cf. Porter, "Experiment Stations in the South," 93; Charles E. Rosenberg, "Science, Technology, and Economic Growth: The Case of the Agricultural Experiment Station Scientist, 1875-1914," *Agricultural History* 45 (Jan. 1971): 1-20; Walter Pittman, "Chemical Regulation in Mississippi: The State Laboratory (1882-)," *Journal of Mississippi History* 41 (May 1979): 133-54; Margaret Ripley Wolfe, "The Agricultural Experiment Station and Food and Drug Control: Another Look at Kentucky Progressivism, 1898-1916," *Filson Club History Quarterly* 49 (Oct. 1975): 323-38.

8. Quotes from Fite, "Southern Agriculture since the Civil War," 14-15; Robert Higgs, "The Boll Weevil, the Cotton Economy, and Black Migration, 1910-1930," *Agricultural History* 50 (Apr. 1976): 335-50; Dunlap, *DDT*, 30.

9. Weevils will eat certain other plants, including hibiscus, hollyhocks, and okra, but rarely. See R.W. Howe, *Studies of the Mexican Boll Weevil in the Mississippi Valley*, USDA Bulletin No. 358 (Washington, D.C., 1916); T.L. Webb, *Cotton or Boll Weevils*, USDA Miscellaneous Publication No. 484 (Washington, D.C., n.d.).

10. Rosenberg, "Science, Technology," 18; quote from Linda O. Hines, "George W. Carver and the Tuskegee Agricultural Experiment Station," *Agricultural History* 53 (Jan. 1979): 71-83.

11. Samuel H. Adams, "Yellow Fever: A Problem Solved, "*McClure's Magazine* 27 (June 1906): 178-92.

12. See J.A. Ferrell and P.A. Mead, *History of County Health Organizations in the United States, 1908-33*, USPH Bulletin No. 222 (Washington, D.C., 1936); quotes from P.F. Russell, "The United States and Malaria: Debits and Credits," *Bulletin of the New York Academy of Medicine* 44 (June 1968): 628, 634.

13. Shryock, *Development of Modern Medicine*, 409-10.

14. Ibid., 415. Information on the hookworm fight, except as otherwise noted, is from John Ettling, *The Germ of Laziness: Rockefeller Philanthropy and Public Health in the New South* (Cambridge, Mass., 1981).

15. Quoted in Frederick Eberson, "Eradication of Hookworm Disease in Florida," *Journal of the Florida Medical Association* 67 (Aug. 1980): 736.

16. Eberson, "Eradication," 741.

17. See George A. Comer, *A History of the Rockefeller Institute, 1901-1953* (New York, 1964), 1-55.

18. Porter, "Experiment Stations," 97; Eberson, "Eradication," 739; William H. Dinsmore, *Hookworm Disease and the Campaign for Its Eradication in Alabama* (Montgomery, 1915); Shirley G. Schoonover, "Alabama Public Health Campaign, 1900-1919," *Alabama Review* 28 (July 1975): 218-33. The anti-hookworm campaign had important consequences for other efforts to improve the health of southern rural children. See William A. Link, "Privies, Progressivism, and Public Schools: Health Reform and Education in the Rural South, 1909-1920," *Journal of Southern History* 54 (Nov. 1988): 623-42.

19. Cf. Faust, "History of Malaria," 124-25. The symptoms of malaria appear to be quite responsive to nutrition. In a well-fed but infected population there is apt to be heavy infant mortality, but adults remain fairly healthy. Among people poorly fed, "chronic malarial ill-health" in adults is much more common. MacFarlane Burnet, *Natural History of Infectious Diseases* (Cambridge, Mass., 1962), 345.

20. An important contributory factor was the lack of doctors in many parts of the South (as in poor and rural areas generally). A 1927 study by the Duke Endowment showed, in the words of Richard Shryock, that "there was one physician for every 772 persons in the United States as a whole, [but] there was only one for every 1,244 in North Carolina, and only one for every 1,409 in South Carolina. In Berkely County, in the latter state, the ratio was as low as one to 4,512." *Development of Modern Medicine*, 415-16.

21. Samuel P. Hays, *Conservation and the Gospel of Efficiency: The Progressive Conservation Movement, 1890-1920* (New York, 1969), especially 1-4.

22. Clepper, *Origins*, 45.

23. Frank B. Vinson, "Conservation and the South, 1890-1920" (Ph.D. diss., University of Georgia, 1971), 42-54; *U.S. Statutes at Large*, 36:961; U.S. Congress, *Senate Document 84*, 57 Cong., 1 sess. (1902).

24. Vinson, "Conservation and the South," 55-93.

25. Ibid., 137; Robert S. Maxwell, "The Impact of Forestry on the Gulf South," *Forest History* 17 (Apr. 1973): 31-33.

26. Quotes from Thomas Gilbert Pearson, *Adventures in Bird Protection* (New York, 1937), 51-52, 74, 82-84; *Journal of the House of Representatives of the General Assembly of the State of North Carolina, Session 1903* (Raleigh, 1903), 137, 277, 426, 1083-84, 1110, 1295.

27. Hynning, *State Conservation*, 92; Vinson, "Conservation and the South," 234.

28. Vinson, "Conservation and the South," 274-75.

29. Doughty, *Feather Fashions*, 14-15, 79-80, 109-10, quote from 155.

30. See Luther J. Carter, *The Florida Experience: Land and Water Policy in a Growth State* (Baltimore, 1974), especially 67-69, 89, 143.

31. Cf. N. Bruce Hannay and Robert E. McGinn, "The Anatomy of Modern Technology: Prolegomenon to an Improved Public Policy for the Social Management of Technology," *Daedalus* 109 (Winter 1980): 27, 44.

32. Huey P. Long, *Every Man a King: The Autobiography of Huey P. Long* (New Orleans, 1933), 42.

33. Hynning, *State Conservation*, 19-20, 98, 100.

34. Ibid., 44.

35. Paragraphs based on: Pete Daniel, *Deep'n As It Come: The 1927 Mississippi River Flood* (New York, 1977), 9; Albert E. Cowdrey, *Land's End* (New Orleans, 1977), 34-37.

36. U.S. Congress, *House Document 798*, 71 Cong., 3 sess. (1931), 1:5.

37. Donald C. Swain, *Federal Conservation Policy, 1921-1933*, University of California Publications in History, No. 76 (Berkeley, 1963). Selections reprinted as "Conservation Accomplishments, 1921-1933," in Nash, *American Environment*, 145.

Chapter 8

1. Basic sources on Engineer work include: *Laws of the United States Relating to the Improvement of Rivers and Harbors* (Washington, D.C., 1940), especially 3:204; and U.S. Congress, *House Document 90*, 70 Cong., 1 sess. (1927).

2. Harold N. Fish, *Geological Investigation of the Atchafalaya Basin and the Problem of Mississippi River Diversion* (Vicksburg, Miss., 1949), 1:8-9, 17-18, 65-68; William Sommer, "Atchafalaya Basin Levee Construction" (Master's thesis, Tulane University, 1966), 8-9; Rodney A. Latimer and Charles W. Schweizer, *The Atchafalaya River Study* (Vicksburg, Miss., 1951), 35.

3. Fisk, *Geological Investigation*, 141.

4. Ferguson recounts his experiments in *History of the Improvement of the Lower Mississippi River for Flood Control and Navigation, 1932-1939* (Vicksburg, Miss., 1940).

5. See Gilbert F. White et al., *Changes in the Urban Occupance of Flood Plains in the United States* (Chicago, 1958).

6. *Annual Report of the Chief of Engineers for 1937* (Washington, D.C., 1938), 1:1678.

7. On Muscle Shoals, see Preston J. Hubbard, *Origins of the TVA* (Nashville, 1961); on the navigational side of TVA, Wilmon H. Droze, *High Dams and Slack Waters: TVA Rebuilds a River* (Baton Rouge, 1965). Also useful are Willis M. Baker, "Reminiscing about the TVA," *American Forests* 75 (May 1965): 31; Norman Wengert, "TVA—Symbol and Reality," *Journal of Politics* 13 (Aug. 1951): 369-92; idem, *Valley of Tomorrow: The TVA and Agriculture* (Knoxville, Tenn., 1952); Clarence Lewis Hodge, *The Tennessee Valley Authority: A National Experiment in Regionalism* (Washington, D.C., 1938); William E. Cole, "The Impact of TVA upon the Southeast," *Social Forces* 28 (May 1950): 435-40.

8. Kenneth J. Seigworth, "Reforestation in the Tennessee Valley," *Public Administration Journal* 8 (Autumn 1948): 280-85.

9. Baker, "Reminiscing about the TVA," 56-60.

10. Useful on this topic is William H. Childress, Jr., "TVA and Cooperative Health Work" (Master's thesis, Syracuse University, 1950).

11. Wengert, *Valley of Tomorrow*, 101.

12. Quotes from R.G. Tugwell and E.C. Banfield, "Grass Roots Democracy—Myth or Reality?" *Public Administration Review* 10 (Winter 1950): 47-55.

13. Lenore Fine and Jesse A. Remington, *Construction in the United States*, U.S. Army in World War II (Washington, D.C., 1972), 683-90.

14. Porter, "Experiment Stations in the South," 95; Floyd W. Hicks and C. Roger Lambert, "Food for the Hungry: Federal Food Programs in Arkansas, 1933-1942," *Arkansas Historical Quarterly* 37 (Spring 1978): 23-43; Paul E. Mertz, *New Deal Policy and Southern Rural Poverty* (Baton Rouge, 1978), 12-14. On pellagra, see Daphne A. Roe, *A Plague of Corn: The Social History of Pellagra* (Ithaca, N.Y., 1973), especially 99-134; Joseph Goldberger, "Pellagra: Causation and a Method of Prevention," *Health in the Southern United States*, 471-76. Quotes on pellagra from Elizabeth W. Etheridge, *The Butterfly Caste: A Social History of Pellagra in the South* (Westport, Conn., 1972), 79, 210.

15. See C.C. Kiker, "Engineering in Malaria Control," *Southern Medical Journal* 34 (Aug. 1941): 839-40; E.L. Bishop, "Cooperative Investigations of the Relation between Mosquito Control and Wildlife Conservation," *Science* 92 (Aug. 30, 1940): 201-2; Justin M. Andrews and Jean L. Grant, "Experience in the United States," in Ebbe Curtis Hoff, ed., *Communicable Diseases: Malaria*, Preventive Medicine in World War II (Washington, D.C., 1963), 6:64-66; Harrison, *Mosquitoes, Malaria and Man*, 218-27.

16. Quote from Mertz, *New Deal Policy*, 2. See also U.S. Department of Commerce, Bureau of the Census, *Mortality Statistics, 1935: Thirty-Sixth Annual Report* (Washington, D.C., 1937): 142-93.

17. Andrews and Grant, "Experience in the U.S.," 75-76, 84, 88-90; Russell, "United States and Malaria," 634; Thomas Parran, "The United States Public Health Service in the War," in Morris Fishbein, ed., *Doctors at War* (New York, 1945), 259-60.

18. Hugh Hammond Bennett, *Soil Conservation* (New York, 1939): 569-72.

19. Much of this section is summarized from John A. Salmon, *The Civilian Conservation Corps, 1933-1942: A New Deal Case Study* (Durham, N.C., 1967).

20. T.B. Plair, "How the CCC Has Paid Off," *American Forests* 60 (Feb. 1954): 28-30, 44-45.

21. Hubert Humphreys, "In a Sense Experimental: The Civilian Conservation Corps in Louisiana," *Louisiana History* 5 (Fall 1964): 345-67; 6 (Winter 1965): 27-52. Quote from 357.

22. Clepper, *Origins*, 50; C.F. Evans, "A Saga of Southern Pine," *American Forests* 48 (Sept. 1945): 403-6, 428; E. Morrell, "Paper Making in the South—A Brief History," *Manufacturers Record* 108 (Apr. 1939): 34-35, 44, 54, 56.

23. W.L. McHale, "The Paper Industry and the Southland Mill," *Journal of Forestry* 50 (July 1952): 536-38; P.M. Garrison, "Building an Industry on Cut-Over Land," *Journal of Forestry* 50 (Mar. 1952): 185-87; T.S. Buie, "From Pines to Pines," *American Forests* 62 (June 1956): 20-23, 54-55; John Solomon Otto, "The Decline of Forest Farming in Southern Appalachia," *Journal of Forest History* 27 (Jan. 1983): 18-27.

24. Ira N. Gabrielson, *Wildlife Refuges* (New York, 1943).

25. Kenneth R. Philp, "Turmoil at Big Cypress: Seminole Deer and the Florida Cattle Tick Controversy," *Florida Historical Quarterly* 56 (July 1977): 28-44; Roosevelt quote on 41. Cf. also Warren W. Chase, "Recent Advances in Forest Game Management," *Journal of Forestry* 47 (Nov. 1949): 882-85.

26. Wellington Brink, *Big Hugh: The Father of Soil Conservation* (New York, 1951), 59. See also Clepper, *Origins*, 96-97. "It is possible," writes Stanley W. Trimble, "to regionalize historical soil erosion in the eastern United States. The dominant pattern ... is that of much more erosion damage below the Mason-Dixon line than above it." Primary causes are rainfall, landform, and farming practices. "Perspectives on the History of Soil Erosion Control in the Eastern United States," *Agricultural History* 59 (Apr. 1985), 162-80. Quote on 170.

27. Bennett, *Soil Conservation*, 9-10; quote from Clepper, *Origins*, 93.

28. Fite, "Southern Agriculture since the Civil War," 16; Public Affairs Committee, *Farm Policies under the New Deal*, Public Affairs Pamphlets No. 16 (1938).

29. See Donald Holley, "Old and New Worlds in the New Deal Resettlement Program: Two Louisiana Projects," *Louisiana History* 11 (Spring 1970): 137-66; Howard Kester, *Revolt among the Sharecroppers* (New York, 1969), 26; Mertz, *New Deal Policy*, 15, 29, 36, 48, 68.

30. Fite, "Southern Agriculture since the Civil War," 19.

31. There were marked exceptions to the general rule of static or declining rural population after 1940 in Virginia, in North Carolina, and especially in Florida. See the tables of rural/urban population by decades in *Statistical Abstract*.

Chapter 9

1. A recent, widely praised, and easily read account of the period is Charles P. Roland, *The Improbable Era: The South Since World War II* (Lexington, Ky., 1975).

2. D. Clayton Brown, "Health of Farm Children in the South, 1900-1950," *Agricultural History* 53 (Jan. 1979): 179-80. See also Elizabeth W. Etheridge, "Pellagra: An Unappreciated Reminder of Southern Distinctiveness," especially 115-16, in Todd L. Savitt and James Harvey Young, *Disease and Distinctiveness in the American South* (Knoxville, Tenn., 1988).

3. Russell, "United States and Malaria," 636-37.

4. Harrison, *Mosquitoes, Malaria and Man*, 236; John Duffy, "The Impact of Malaria on the South," in Savitt and Young, *Disease and Distinctiveness in the American South*, 49.

5. Brown, "Health of Farm Children," 180-81; Mertz, *New Deal Policy*, 255.

6. Fite, "Southern Agriculture since the Civil War," 4.

7. Harry D. Fornari, "The Big Change: Cotton to Soybeans," *Agricultural History* 53 (Jan. 1979): 245-53.

8. Fite, "Southern Agriculture since the Civil War," 4, 18.

9. C.H. Blackman, "The Tung-Oil Industry," *Botanical Review* 9 (Jan. 1943): 1.

10. Quotes from Fite, "Southern Agriculture since the Civil War," 19; and Roland, *Improbable Era*, 21.

11. Quotes from Dunlap, *DDT*, 73; J.H. Perkins, "Boll Weevil Eradication," *Science* 207 (Mar. 7, 1980): 1044-51; and Douglas Helms, "Technological Methods for Boll Weevil Control," *Agricultural History* 53 (Jan. 1979): 295, 297. See also G.M. Woodwell, C.F. Wurster, and P.A. Isaacson, "DDT Residues in an East Coast Estuary: A Case of Biological Concentration of a Resistant Pesticide," *Science* 156 (1967): 821-24.

12. Quote from Dunlap, *DDT*, 78. On the worldwide attack upon malaria, see Harrison, *Mosquitoes, Malaria and Man*, 208-60. On Florida eagles, see Charles L. Broley, "The Plight of the American Bald Eagle," *Audubon Magazine* 60 (July-Aug. 1958): 162.

13. S.H. Spurr, "Progress in Silviculture," in Henry Clepper and Arthur B. Meyer, eds., *American Forestry: Six Decades of Growth* (Washington, D.C., 1960), 72; Paul F. Sharp, "The Tree-Farm Movement: Its Origin and Development," *Agricultural History* 23 (Jan. 1949): 43; Clepper, *Origins*, 51-53; Hunt, *Physiography*, 137.

14. Edmund H. Fulling, "Botanical Aspects of the Paper-Pulp and Tanning Industries in the United States—An Economic and Historical Survey," *American Journal of Botany* 43 (Oct. 1956): 621-25; Helen Hunter, "Innovation, Competition, and Locational Changes in the Pulp and Paper Industry, 1880-1950," *Land Economics* 31 (Nov. 1955): 314-27; Merle C. Prunty, Jr., "Recent Expansions in the Southern Pulp-Paper Industries," *Economic Geography* 32 (Jan. 1956): 51-57; Reuben B. Robertson, "Recent Developments in Southern Forestry," *Georgia Review* 5 (Fall 1951): 362-68.

15. Ignatz James Pikl, *A History of Georgia Forestry* (Athens, Ga., 1966); J.M. Stouffer, "Forestry in Alabama," *Alabama Historical Quarterly* 10 (1948): 65-66.

16. V.L. Harper, "Forestry Research"; C. Raymond Clar, "State Forestry"; and George

A. Garratt, "Six Decades of Growth," are all in Clepper and Meyer, *Six Decades*, 36-49, 209-24, and 1-23, respectively. See also Elbert H. Reid and Raymond Price, "Progress in Forest Range Management," also in Clepper, *Six Decades*, 112-22. On fire, see Stephen J. Pyne, *Fire in America: A Cultural History of Wildland and Rural Fire* (Princeton, 1982), especially 159. For an excellent analysis of the forest industry's growth and structure and its impact on a very poor state, see Warren A. Flick, "The Wood Dealer System in Mississippi: An Essay on Regional Economics and Culture," *Journal of Forest History* 29 (July 1985): 131-38.

17. Oliver A. Houck, "The Regulation of Toxic Pollutants Under the Clean Water Act," *Environmental Law Reporter* 21 (Sept. 1991): 10551-10553. Quote from "The View from Hilton Head," *Harper's Magazine* 240 (May 1970): 103. See also James M. Fallows et al., *The Water Lords* (New York, 1971).

18. On timber companies, see Milton Moskowitz, Michael Katz, and Robert Levering, *Everybody's Business, An Almanac: The Irreverent Guide to Corporate America* (New York, 1980), 567-90. On the "carbon sink" claim, see Hazel R. Delacourt and W.F. Harris, "Carbon Budget of the Southeastern U.S. Biota: Analysis of Historical Change in Trend from Source to Sink," *Science* 210 (Oct.17, 1980): 321-22. See also Fred C. Simmons, "Not Seeing the Trees for the Forest," *New York Times*, Jan. 15, 1981. Southern acreage in forestland rose in 1945-1960 and thereafter declined slightly; volume of both saw timber and growing stock increased steadily. See USDA's annual *Agricultural Statistics* for details.

19. Southeastern Association of Fish and Wildlife Agencies, *Proceedings of the Annual Conference, 1978* (n.p., n.d.), 546. This source will hereafter be cited as *SE Wildlife Conference Proceedings* with the relevant date.

20. *SE Wildlife Conference Proceedings, 1978*, 82, 150-51, 182-85, 266-67; *1975*, 441-50. See also *New York Times*, Jan. 19 and 31, 1981; Matthiesen, *Wildlife*, 50-51; *Federal Register* 45 (May 20, 1980): 33769-81; Robert Fritchey, "Plight of the Pelican," *Louisiana Life* (Autumn 1994): 45-49.

21. *SE Wildlife Conference Proceedings, 1989*, 5, 8.

22. Ibid., 8.

23. Clepper, *Origins*, 78-79; Hunt, *Physiography*, 165. On the fire ant's entry, see *SE Wildlife Conference Proceedings, 1976*, 414; on the armadillo, see *SE Wildlife Conference Proceedings, 1977*, 57.

24. Raymond F. Dasmann et al., *Environmental Impact of the Cross-Florida Barge Canal with Special Emphasis on the Oklawaha Regional Ecosystem* (Gainesville, Fla., 1971), 81, 83. See also U.S. Congress, *House Document 37*, 85 Cong., 1 sess. (1956), 1-3.

25. See Samuel P. Hays, *Beauty, Health, and Permanence: Environmental Politics in the United States, 1955-1985* (Cambridge, Mass., and New York, 1987), especially 2-10, 521-26.

26. Albert E. Cowdrey, "Pioneering Environmental Law: Army Engineers and the Refuse Act," *Pacific Historical Review* 44 (Aug. 1975): 344. On sectional behavior in voting on environmental issues, see annual reports by the League of Conservation Voters, Washington, D.C. On environmental laws in general, see Mary Robinson Sive, ed., *Environmental Legislation: A Sourcebook* (New York, 1976). Quote in Cassandra Tate, "Ambivalent TVA Roles in Energy and Conservation," *Smithsonian* 10 (Jan. 1980): 94-95.

27. Peter Matthiesen, "How to Kill a Valley," *New York Review of Books* 27 (Feb. 7, 1980): 32. Description of the snail darter from *Hill* v. *Tennessee Valley Authority*, 549 F.2d 1064, 9 ERC 1737 (6th Cir. 1977), in Oscar S. Gray, *Cases and Materials on Environmental Law*, 1977 Supplement, 2nd ed. (Washington, D.C., 1977), 156.

28. Gray, *Cases and Materials*, 158-59; "Skeptical Eye," *Discover* 2 (Feb. 1981): 14.

29. Other sources of this section include: Sara Grigsby Cook, Chuck Cook, and Doris Gove, "What They Didn't Tell You about the Snail Darter and the Dam," *National Parks and Conservation Magazine* 51 (May 1977): 10-13; James Branscombe, "The TVA: It Ain't What It Used to Be," *American Heritage* 28 (Feb. 1977): 68-78; John Dernbach, " 'Little

Fish' Versus 'Big Dam,'" *Progressive* 42 (Dec. 1978): 45-48. On August 28, 1980, the *Washington Post* reported the Cherokees' failure to appeal their suit.

30. Jeffrey K. Stine, *Mixing the Waters: Environment, Politics, and the Building of the Tennessee-Tombigbee Waterway* (Akron, Ohio, 1993) provides the best account of Tenn-Tom to date. Most other printed accounts have been adversary writing—newspaper stories drawn from railroad handouts, or puffery by supporters. A vigorous attack is James Nathan Miller, "Trickery on the Tenn-Tom," *Reader's Digest* (Sept. 1978): 138-43.

31. Merrens, *Colonial North Carolina*, 180; J.S. Paterson, *North America: A Geography of Canada and the United States* (New York, 1975), 230; Hunt, *Physiography*, 159-60; *Wall Street Journal*, Sept. 23, 1994; Merrill L. Johnson, "Postwar Industrial Development in the Southeast and the Pioneer Role of Labor-Intensive Industry," *Economic Geography* 61 (Jan. 1985): 46-65.

32. *New York Times*, July 31, 1994; Oliver A. Houck, "This Side of Heresy: Conditioning Louisiana's Ten-Year Industrial Tax Exemptions upon Compliance with Environmental Laws," *Tulane Law Review* 61 (Dec. 1986): 289-378. On the shift of industry toward the sunbelt, see, e.g., Emilio Cassetti, "Manufacturing Productivity and Snowbelt-Sunbelt Shifts," *Economic Geography* 60 (Oct. 1984): 313-23 and sources cited therein.

33. S.B. Cannon et al., "Epidemic Kepone Poisoning in Chemical Workers," *American Journal of Epidemiology* 107 (1978): 529-37; quote from 531. See also *New York Times*, Aug. 30, Dec. 25, Dec. 27, 1975.

34. *Atlanta Constitution*, March 7, 1994; *New Orleans Times-Picayune*, Nov. 3 and 23, 1994; Jan Bowermaster, "A Town Called Morrisonville," *Audubon* 95 (Jul.-Aug. 1993): 42-51. Author interviews with Oliver A. Houck, Tulane Law School, Sept. 2, 1994; Mark Schleifstein, *Times-Picayune* environmental correspondent, by telephone, Oct. 26, 1994; Victor Alexander, M.D., occupational medicine specialist, by telephone, Oct. 27, 1994; and William Fontenot, Louisiana Attorney General's office, by telephone, Nov. 4, 1994. See also Houck, "Regulation of Toxic Pollutants Under the Clean Water Act," 10556; Senator Albert Gore, *Earth in the Balance: Ecology and the Human Spirit* (Boston, 1992), 290-91; *Atlas of U.S. Cancer Mortality among Whites: 1950-1980*, DHHS Publication No. (NIH) 67.2900 (Washington, D.C., n.d.), 24-25; *Atlas of U.S. Cancer Mortality among Nonwhites: 1950-1980*, NIH Publication No. 90-1582 (Washington, D.C., n.d.), 26-27. LSU Medical Center is now engaged in an EPA-funded study to determine the effect of industrial pollution on cancer in the Louisiana industrial corridor.

35. Branley Allan Branson, "Strip Mining and the Environment," *National Parks and Conservation Magazine* 51 (Apr. 1977): 10-12; *Federal Register*, vol. 59, No. 103 (May 31, 1994): 27989-28013.

36. Hunt, *Physiography*, 159. See report in *Science* 195 (1977): 137.

37. Frank Graham, Jr., "Oil in the Sea: How Little We Know," *Audubon* 80 (Nov. 1978): 133, 137-40, 144-46; C.R. Grau, T. Rondybush, J. Dobbs, and J. Wather, "Altered Yolk Structure and Reduced Hatchability of Eggs from Birds Fed Single Doses of Petroleum Oils," *Science* 195 (Feb. 25, 1977): 779-81. On waterbird poisoning, see *SE Wildlife Conference Proceedings, 1977*, 180-87; *1978*, 309-25.

38. Gladwin Hill, "Ixtoc's Oil Has a Silver Lining," *Audubon* 81 (Nov. 1979). Quotes from 150.

39. Paul Kalman, "Oil and Water: Can They Mix?" *Field and Stream* 79 (Mar. 1975): 149.

40. Paragraphs based on: James K. Mitchell, *Community Response to Coastal Erosion* (Chicago, 1974); Robert Dolan and Bruce Hayden, "Adjusting to Nature in Our National Seashores," *National Parks and Conservation Magazine* 48 (June 1974): 9-14; Vincent Bellis, Michael P. O'Connor, and Stanley Riggs, *Estuarine Shoreline Erosion in the Albemarle-Pamlico Region of North Carolina* (Greenville, N.C., 1975). Quote from Mitchell, *Coastal Erosion*, 58.

41. Carter, *Florida Experience*, especially 57-264; Art Marshall, "Alligators and Cities: Lessons from the Everglades," *Florida Naturalist* 45 (Aug. 1972): 107-9; George Reiger,

"The Choice for Big Cypress: Bulldozers or Butterflies," *National Wildlife* 10 (Oct.-Nov. 1972): 4-11; and idem, "The River of Grass Is Drying Up," *National Wildlife* 12 (Dec.-Jan. 1974): 54-62; Mark Derr, "Redeeming the Everglades," *Audubon* 95 (Sept.-Oct. 1993): 48-56, 128-31.

42. On recent Florida politics, see Jack Bass and Walter DeVries, *The Transformation of Southern Politics: Social Change and Political Consequence since 1945* (New York, 1977), 107-35. For a summary on the jetport settlement, see Gray, *Cases and Materials,* 569. Quotes from *SE Wildlife Conference Proceedings, 1989,* 550-51.

43. *New York Times,* Mar. 22, 1994 (quote), and May 4, 1994; *Annual Report of the Chief of Engineers, 1992* (Washington, D.C., 1994), 9-25; Derr, "Redeeming the Everglades," 52; Vincent Kiernan, "Refreshment for the Thirsty Everglades," *New Scientist* 139 (July 24, 1993): 10. Groundbreaking for the Kissimmee project took place on April 23, 1994, according to Jacquelyn Griffin, chief of public affairs for the Jacksonville District, Corps of Engineers. See the district's publication *Central and Southern Florida Project Review Study News* (June and Oct. 1994) for an account of the larger project plans and development process.

44. Philip Bowman, Dr. Ben Lahr, and Paul Coreil, "Louisiana's Vanishing Coast: Processes and Problems," Pt. III, *Louisiana Conservationist* (May/June 1994): 4-6; Environmental Protection Agency, *America's Wetlands: Our Vital Link Between Land and Water* (n.p., n.d.); Louisiana Wetlands Conservation and Restoration Task Force, *Louisiana Coastal Wetlands Restoration Plan* (Nov. 1993), especially 7, 25-40.

45. On work currently under way in the basin/floodway, see U.S. Army Corps of Engineers, *Water Resources Development in Louisiana, 1993* (New Orleans, 1993), 28-43. *Atchafalaya Basin* (n.p., n.d.), a brochure issued jointly by the federal agencies and the state and available from the EPA and the Corps of Engineers, sums up the existing problem and underlines the current lack of a multipurpose plan to harmonize the diverse interests struggling in the basin.

46. This section is based primarily on: interview with Oliver A. Houck, National Wildlife Federation, Washington, D.C., Apr. 17, 1981; James H. Carmichael, Jr., "It's Not Too Late to Save the Atchafalaya," *National Wildlife* 10 (Apr.-May 1972): 26-30; Grits Gresham, "Atchafalaya Basin Crisis," *Louisiana Conservationist* 15 (Nov.-Dec. 1963): 10-11; Charles Fryling, Jr., "Atchafalaya: America's Largest River Basin Swamp and Unique Delta Treasure," *Ozark Society Bulletin* 12 (Autumn 1978): 3-7; R.D. Adams et al., *Shoreline Erosion in Coastal Louisiana: Inventory and Assessment* (Baton Rouge, 1978); Cowdrey, "Pioneering Environmental Law"; J.D. Lee, "Wetland Litigation in the Gulf Coast States," *Mississippi Law Journal* 58 (Spring 1988): 123-54; Jeffrey K. Stine, "Regulating Wetlands in the 1970s: U.S. Army Corps of Engineers and the Environmental Organizations," *Journal of Forest History* 27 (Apr. 1983): 60-75 (quote, 61). The Clean Water Act is administered jointly by the Corps of Engineers and the Environmental Protection Agency.

47. The law is the Coastal Wetlands Planning, Protection and Restoration Act. See Robert D. Brown, "Restoring Louisiana's Wetlands," *Military Engineer* No. 550 (July 1992): 2-7 (quote, 4). Precise numbers for the different parts of the coastal plain can be found in Louis D. Britsch and E. Burton Kemp III, *Land Loss Rates,* U.S. Army Corps of Engineers Technical Report GL-90-2 (Apr. 1990), 3 vols., available from the Waterways Experiment Station, Vicksburg, Miss. Quote from Mark Schleifstein, "Saving the Coast," New Orleans *Times-Picayune,* May 8, 1994.

Chapter 10

1. Quote from Cooke, *Vintage Mencken,* 199; Charles P. Roland, "The South, America's Will-o'-the-Wisp-Eden," *Louisiana History* 11 (Spring 1970): 101-20.

2. Malcolm Cowley, ed., *The Portable Faulkner* (New York, 1977), xxv-xxvi.
3. Marjorie Kinnan Rawlings, *Cross Creek* (New York, 1942), 263, 368.
4. Hearn, *Chita*, 5, 146, 163.
5. Cowley, *Portable Faulkner*, 197, 200, 318.
6. Redding S. Sugg, Jr., ed., *The Horn Island Logs of Walter Anderson* (Memphis, 1973), 19-20, 23, 32. Biographical details from Sugg's introduction.
7. Aldo Leopold, "The Conservation Ethic," *Journal of Forestry* 31 (1933): 635-38; Joseph M. Petulla, *American Environmentalism: Values, Tactics, Priorities* (College Station, Tex., 1980), especially 204-31.

Bibliographical Note

To the best of my knowledge, no previous work covers the entire field addressed by the present study. The works of the regionalist Howard W. Odum and historical geographers like Ralph H. Brown probably come closest. Acknowledged classics in southern and in environmental history, by C. Vann Woodward, Clement Eaton, Ulrich B. Phillips, Samuel P. Hays, Roderick Nash, and Aldo Leopold, are relevant to all work in their respective fields but will not be listed here. Among more specialized works that I found of most value were the following:

On the basic physical structure of the region, Charles B. Hunt, *Physiography of the United States* (San Francisco, 1969), was exceptionally helpful. The southern climate, overemphasized by the climatic determinists of the past, received renewed scholarly attention in A. Cash Koeniger, "Climate and Southern Distinctiveness," *Journal of Southern History* 54 (Feb. 1988): 21-44. He seems to have found both reliable data and a useful method of approaching this much-disputed topic. Southeastern Indians have received much scholarly attention from the ethnographers of the late nineteenth and early twentieth centuries. Among recent books, I found Charles Hudson, *The Southeastern Indians* (n.p., 1976), especially illuminating, and Gordon R. Willey, *An Introduction to American Archeology: North and Middle America* (Englewood Cliffs, N.J., 1966), a useful survey. Francis Jennings, *The Invasion of America: Indians, Colonization, and the Cant of Conquest* (Chapel Hill, 1975), demonstrates that a blaze of indignation can sometimes generate light. Calvin Martin, *Keepers of the Game: Indian Animal Relationship and the Fur Trade* (Berkeley, 1978), though not concerned with the region, is a most original and intriguing essay. Another study of northern Indians that well repays reading is Gordon M. Day, "The Indian as an Ecological Factor in the Northeastern Forest," *Ecology* 34 (1953): 329-46. J. Leitch Wright, Jr., *The Only Land They Knew: The Tragic Story of the American Indians in the Old South* (New York, 1981), is a fine study.

The reader interested in the early years of European settlement will find much to enjoy as well as to use in David B. Quinn, *North America from Earliest Discovery to First Settlements: The Norse Voyages to 1612* (New York, 1977), and especially Quinn's *The Roanoke Voyages* (Cambridge, Mass., 1955). Citations to some other basic works on the early explorations are found in the notes. St. Julian R. Childs, *Malaria and Colonization in the Carolina Low Country, 1526-1696* (Baltimore, 1940), and Wyndham B. Blanton, *Medicine in Virginia in the Seventeenth Century* (Richmond, 1930), are still of interest for anyone concerned with the transformation of the disease environment. Informative on Louisiana during settlement is the first volume of John Duffy's *The Rudolph Matas History of Medicine in Louisiana* (Baton Rouge, 1958). The same author's *Epidemics in Colonial America* (Baton Rouge, 1953) is a standard source. Patterns of agriculture are explored in Avery Craven's classic *Soil Exhaustion as a Factor in the Agricultural History of Virginia and Maryland, 1606-1860* (1926; reprint, Gloucester, Mass., 1965), which retains its freshness as its seventieth anniversary approaches, while Louis Cecil Gray, *History of Agriculture in the Southern United States to 1860* (1932; reprint, Clifton, N.J., 1973), after a life almost as long, seems if anything to grow in influence.

In the last few decades, a number of exciting works have thrown unexpected light upon the ecology of settlement. Of exceptional interest in obtaining an overview of the whole process in the New World is Alfred W. Crosby, *The Columbian Exchange: Biological and Cultural Consequences of 1492* (Westport, Conn., 1972). T.D. Stewart presents the "cold screen" theory in "A Physical Anthropologist's View of the Peopling of the New World," *Southwestern Journal of Anthropology* 16 (Autumn 1960): 259-73. Carville V. Earle has produced two justly celebrated studies of the early Chesapeake in *The Evolution of a Tidewater Settlement System: All Hallow's Parish, Maryland, 1650-1783* (Chicago, 1975), as well as "Environment, Disease, and Mortality in Early Virginia," in Thad W. Tate and David L. Ammerman, eds., *The Chesapeake in the Seventeenth Century: Essays on Anglo-American Society* (Chapel Hill, 1979), 96-125. Darrett B. Rutman and Anita N. Rutman add to our knowledge of disease patterns in "Of Agues and Fevers: Malaria in the Early Chesapeake," *William and Mary Quarterly* 33 (1976): 31-60, while H. Roy Merrens and George D. Terry review the course of events in a Deep South colony in "Dying in Paradise: Malaria, Mortality, and the Perceptual Environment in Colonial South Carolina," *Journal of Southern History* 50 (Nov. 1984): 533-50. Allan L. Kulikoff, *Tobacco and Slaves: The Development of Southern Culture in the Chesapeake, 1680-1800* (Chapel Hill, 1986), and Daniel B. Smith, "Mortality and Family in the Colonial Chesapeake," *Journal of Interdisciplinary History* 8 (Winter 1978): 403-27, present remarkable views of the impact of a high death rate on the Chesapeake society.

The literature of slavery is stunning in extent. Broad, compact, and richly documented is Ira Berlin's essay "Time, Space, and the Evolution of Afro-American Society on British Mainland North America," *American Historical Review* 85 (Feb. 1980): 44-78. Philip D. Curtin, "Epidemiology and the Slave Trade," *Political Science Quarterly* 83 (June 1968): 190-216, is seminal; Joseph C. Miller explores the central black trauma in "Mortality in the Atlantic Slave Trade: Statistical Evidence on Causality," *Journal of Interdisciplinary History* 11 (Winter 1981): 385-424; and Todd L. Savitt writes more broadly than his title would indicate in *Medicine and Slavery: The Diseases and Health Care of Blacks in Antebellum Virginia* (Urbana, Ill., 1978).

Anyone reading these studies might well conclude that the attitude of historians to disease, as to other physical constraints, appears to be undergoing a marked change. When I began to study history seriously in the 1950s, cultural determinism was rampant; indeed, culture was a sort of unmoved mover, the cause of all things but itself uncaused by anything, except by culture. The newer view, surely one buttressed by experience, is that disease often represents a fundamental limiting factor in cultural as in individual life.

On wildlife and forests, especially in the colonial period, I have used and recommended as excellent sources on legal aspects Thomas A. Lund, "British Wildlife Law before the American Revolution," *Michigan Law Review* 74 (Nov. 1975): 49-74; and idem, *American Wildlife Law* (Berkeley, 1980). Arlie W. Schorger has written with scholarship and scientific knowledge on *The Passenger Pigeon: Its Natural History and Extinction* (Madison, Wis., 1955) and on *The Wild Turkey: Its History and Domestication* (Norman, Okla., 1966). Walter P. Taylor has performed a similar service as editor of *The Deer of North America ... Their History and Management* (Washington, D.C., 1956). More recently, Mikko Saikku explored the fate of another lost species in "The Extinction of the Carolina Parakeet," *Environmental History Review* 14 (Fall 1990): 1-20. The literature of the fur trade needs no rehearsing here. Much useful material on the economy of the backwoods can be derived from Richard G. Lillard's literary and ever readable *The Great Forest* (New York, 1947) and from Henry H. Merrens's careful, scholarly *Colonial North Carolina in the Eighteenth Century: A Study in Historical Geography* (Chapel Hill, 1964). I have ventured to quote at some length the observations of an eighteenth century settler, Johann Martin Bolzius, as recorded in his "Reliable Answer to Some Submitted Questions Concerning the Land Carolina, in Which Answer, However, Regard Is Also Paid to the Condition of the Colony of Georgia," trans. and ed. by Klaus G. Loewald, Beverly Starika, and Paul S. Taylor, *William and Mary Quarterly*, 3d ser., 14 (Apr. 1957): 223-61. More recently the same intriguing history has been made available in book form: George F. Jones and Don Savelle, eds., *Detailed Reports on the Salzburger Emigrants Who Settled in America* ... (Athens, Ga., 1983). Traditional sources of information (Jay P. Kinney on forest legislation and Samuel Trask Dana on forest and range policy, plus a variety of compilations of colonial and state laws) are indicated in the notes to chapter 3.

Finally, two large works, difficult to read but incomparable for browsing, must be mentioned. For those who can endure its writer's disposition, Jack P. Green, ed., *The Diary of Colonel Landon Carter of Sabine Hall, 1752-1778* (Charlottesville, Va., 1965), is a treasure of exact information on the life and farming of a great landowner. Richard Beale Davis's prizewinning, encyclopedic *Intellectual Life in the Colonial South, 1585-1763* (Knoxville, Tenn., 1978) is a splendid source somewhat deficient in interpretation, a dictionary of southern intellectual biography for the period.

The nineteenth-century South presents so broad an array of sources on environmental history that any overview must be too brief. Among the travelers' narratives, I found Thomas Nuttall, *Journal of Travels in the Arkansas Territory during the Year 1819* (Philadelphia, 1821), full of interest. The contemporary assumption that environment caused disease shaped the work of medical writers. Daniel Drake, *A Systematic Treatise, Historical, Etiological, and Practical of the Principal Diseases of the Interior Valley of North America* (1850-1854; reprint,

New York, 1971), has attained classic status, and the various publications of Josiah C. Nott, some of which are cited in the notes to chapter 5, are, if possible, of even greater relevance. The many fine works on specific epidemics and diseases, including Charles E. Rosenberg on cholera and John Duffy on the New Orleans yellow fever epidemic of 1853, are well known. Less familiar sources that address the social impact of disease with insight are the following four articles: John H. Ellis, "Business and Public Health in the Urban South during the Nineteenth Century: New Orleans, Memphis, and Atlanta," *Bulletin of the History of Medicine* 44 (May-June 1970): 197-212, (July-Aug. 1970): 346-77; David R. Goldfield, "The Business of Health Planning: Disease Prevention in the Old South," *Journal of Southern History* 42 (Nov. 1976): 557-70; JoAnn Carrigan, "Impact of Epidemic Yellow Fever on Life in Louisiana," *Louisiana History* 4 (Winter 1963): 5-34; and idem, "Privilege, Prejudice, and the Strangers' Disease in Nineteenth Century New Orleans," *Journal of Southern History* 36 (Nov. 1970): 568-78. Recent books by several of the same authors, reviewing and expanding their earlier work, are listed below.

Here, however, I would like to note the importance to this study generally of local historical journals. Not only in Louisiana but throughout the South such publications record much of what has happened to local resources, including the human. On a regional level, the *Journal of Southern History* has often been hospitable to studies of disease. For those needing a nineteenth-century source, *DeBow's Review*, by reason of its economic slant, has much to offer.

Agriculture in the South is a topic to make the nonspecialist know fear. If its bibliography does not yet outstrip the Civil War—plowshare not yet mightier than the sword—it surely equals or surpasses that wordy epic in the sophistication and fine-spunness of its arguments. Nevertheless, agriculture *was* the South, economically and in environmental impact up until the Civil War, and remained a major influence thereafter. Studies of the agronomy and history of cotton, tobacco, and sugarcane are given in the notes. Sam B. Hilliard, *Hog Meat and Hoecake: Food Supply in the Old South* (Carbondale, Ill., 1972), treats some wild as well as cultivated sources of nourishment. John Alfred Heitman, *The Modernization of the Louisiana Sugar Industry, 1830-1910* (Baton Rouge, 1987), provides a model account of the evolution of an important agri-industry in its home state. Julius Rubin, "The Limits of Agricultural Progress in the Nineteenth-Century South," *Agricultural History* 49 (Apr. 1975): 362-74, presents an overall view of how southern agriculture interacted with its natural matrix. Richard C. Sheridan, "Chemical Fertilizers in Southern Agriculture," *Agricultural History* 53 (Jan. 1979): 308-18, points up the enduring significance of this technological "fix" to southern growers. In the post-Civil War period, and extending into the present, Gilbert C. Fite, "Southern Agriculture since the Civil War: An Overview," *Agricultural History* 53 (Jan. 1979): 3-21, presents in a few pages the distillation of much scholarship and thought. Of note on the question of erosion is Stanley W. Trimble, *Man-Induced Soil Erosion on the Southern Piedmont, 1700-1970* (Ankeny, Iowa, 1974), which appears to provide a quantitative guide to this much-discussed topic. His thesis is further developed in "Perspectives on the History of Soil Erosion Control in the Eastern United States," *Agricultural History* 59 (Apr. 1985): 162-80, which points to solid evidence on the regionalization of soil erosion. Robert W. Fogel and Stanley L.

Engerman, *Time on the Cross: The Economics of American Negro Slavery* (Boston, 1974), 1:167-68, have provided some useful asides on difficulties imposed by the regional climate and resource endowment. Engerman and Harold D. Woodman present intriguing views of regional progress in "A Reconsideration of Southern Economic Growth, 1770-1860" and "New Perspectives on Southern Economic Development: A Comment," both in *Agricultural History* 49 (Apr. 1975): 343-61, 374-80. References to a number of other works, special and general, are found in the notes.

On public land policy, a useful introduction may be gleaned from Paul Wallace Gates, "Federal Land Policy in the South, 1866-1888," *Journal of Southern History* 6 (Aug. 1940): 303-30, and idem, "Federal Land Policies in the Southern Public Land States," *Agricultural History* 53 (Jan. 1979): 206-27, as well as Warren Hoffnagle, "The Southern Homestead Act: Its Origin and Operation," *Historian* 32 (Aug. 1970): 612-29. The notes to chapter 6 also contain a number of studies of particular subregions and industries that made use of forests and forestland.

On the development of the Mississippi River, the literature of the century is abundant; during the 1870s in particular, the question took on the aura of a major national endeavor, as in fact it was. Mark Twain's classic *Life on the Mississippi* is not likely to be supplanted as an introduction to the perils and splendors of navigation on the unimproved river. Three works by leading engineers who devised different programs for taming the great river are: Andrew A. Humphreys and Henry L. Abbot, *Report upon the Physics and Hydraulics of the Mississippi River* (Washington, D.C., 1861, 1876); Charles Ellet, *The Mississippi and Ohio Rivers: Containing Plans for the Protection of the Delta* (Philadelphia, 1853); and James B. Eads, *Physics and Hydraulics of the Mississippi River: Report of the U. S. Levee Commission Reviewed* (New Orleans, 1876). Humphreys and Abbot draw a bead on Ellet, as Eads does on them; yet Ellet's book is the most prophetic of the three. Anyone interested in the topic will also wish to consult the *Annual Reports of the Chief of Engineers* and the *Reports* and *Proceedings of the Mississippi River Commission*.

Southern wildlife has received notice in several enjoyable and instructive books since at least the time of William Byrd II. In the antebellum South, William Elliot's *Carolina Sports by Land and Water* (New York, 1967) exhibits not only the joy of the outdoors but also the attitude of the gentleman sportsman who was to be so important to the conservation cause at a later date. Thomas Gilbert Pearson, *Adventures in Bird Protection* (New York, 1937), reveals the southern as well as the national roots of one form of conservation. One major issue in the early conservation movement has received lively treatment in Robin W. Doughty, *Feather Fashions and Bird Preservation: A Study in Nature Protection* (Berkeley, 1975), and the same author has examined the fate of wildlife in a pioneer subculture in *Wildlife and Man in Texas: Environmental Change and Conservation* (College Station, Tex., 1983). Though this bibliographical note is largely limited to published works, it would be difficult to omit Frank B. Vinson, "Conservation and the South, 1890-1920" (Ph.D. diss., University of Georgia, 1971). A full treatment of the political aspects of Progressive conservation in the South would be welcome, however.

The interest of scholars in the major southern diseases has continued to produce works of merit. Margaret Humphreys, *Yellow Fever and the South* (New Brunswick, N.J., 1992), provides a history of the region's most famous illness that is concise, comprehensive, and deeply researched. John Ellis reviews the effects of the epidemic of 1878 in his detailed and well-written *Yellow Fever & Public Health in the New South* (Lexington, Ky., 1992). Older studies of the epidemic by Gerald M. Capers, "Yellow Fever in Memphis in the 1870's," *Mississippi Valley Historical Review* 24 (1938): 483-502, and Thomas H. Baker, "Yellowjack: The Yellow Fever Epidemic of 1878 in Memphis, Tennessee," *Bulletin of the History of Medicine* 42 (May-June 1968): 241-64, are still of interest. Gordon Harrison's excellent *Mosquitoes, Malaria and Man: A History of the Hostilities since 1880* (New York, 1978) has much to offer a historian of the South. On pellagra, we are fortunate to possess in Elizabeth W. Etheridge, *The Butterfly Caste: A Social History of Pellagra in the South* (Westport, Conn., 1972), and Daphne A. Roe, *A Plague of Corn: The Social History of Pellagra* (Ithaca, N.Y., 1973), two examples of what the informed scholar can achieve using the history of disease to throw light upon culture. To the many older studies of hookworm, John Ettling has added a prizewinning contribution in *The Germ of Laziness: Rockefeller Philanthropy and Public Health in the New South* (Cambridge, Mass., 1981). Here the question addressed is, rather, why and how scientists and philanthropists do battle with disease, but the result is equally enlightening. The question of whether southern disease has formed an element in the region's peculiarity is addressed by many of the experts in the field in Todd L. Savitt and James Harvey Young, *Disease and Distinctiveness in the American South* (Knoxville, Tenn., 1988).

The recent history of the southern environment is voluminous but fragmentary, making the selection of works with some resonance beyond their immediate topic more difficult than it might appear. Mississippi River development still requires study of the Corps of Engineers reports cited earlier; their literary value declines precipitously after about 1920. Harley B. Ferguson, *History of the Improvement of the Lower Mississippi River for Flood Control and Navigation, 1932-1939* (Vicksburg, Miss., 1940), is still important, and district histories are available from the engineers at the southern district and division offices. Albert E. Cowdrey, "Pioneering Environmental Law: The Army Corps of Engineers and the Refuse Act," *Pacific Historical Review* 44 (Aug. 1975), 331-49, takes a basic clean-water law across the historical divide from navigational to environmental policy. TVA is the subject of an extensive body of literature, with Wilmon F. Droze, *High Dams and Slack Waters: TVA Rebuilds a River* (Baton Rouge, 1965), an excellent study of navigation. The essay by Willis M. Baker, "Reminiscing about the TVA," *American Forests* 75 (May 1965): 31, 56-60, was particularly useful in this study, as was Kenneth J. Seigworth, "Reforestation in the Tennessee Valley," *Public Administration Journal* 8 (Autumn 1948): 280-85. In the absence of the definitive biography, which is still lacking, the sources of the soil conservation story can be sought in Hugh Hammond Bennett, *Soil Conservation* (New York, 1939), and Wellington Brink, *Big Hugh: The Father of Soil Conservation* (New York, 1951). Besides the local studies indicated in the notes, John A. Salmon, *The Civilian Conservation Corps, 1933-1942: A New Deal Case Study* (Durham, N.C., 1967), gives the general history of the agency. An article with implications beyond its title is D. Clayton Brown, "Health of Farm Children in the South, 1900-1950," *Agricultural History* 53 (Jan. 1979): 170-87. In the same issue, a fundamental transformation in the rural South is treated in Harry D. Fornari, "The Big Change: Cotton to Soybeans," 245-53. Charles P. Roland's *The Improbable Era: The South Since World War II* (Lexington, Ky., 1975) provides a concise account of the most recent period. Thomas R. Dunlap, *DDT: Scientists, Citizens and Public Policy* (Princeton, 1981), gives an enlightening account of how pesticide control evolved before and since the publication of Rachel Carson's *Silent Spring*.

Environmentalism has developed its own vast literature. Since the original publication of this book, Samuel P. Hays, *Beauty, Health & Permanence: Environmental Politics in the United States, 1955-1985* (Cambridge, Mass., and New York, 1987), has provided the most comprehensive and subtle account to date of the politics of the movement through the early years of the Reagan counterrevolution. The Smithsonian's Jeffrey K. Stine, in *Mixing the Waters: Environment, Politics, and the Building of the Tennessee-Tombigbee Waterway* (Akron, Ohio, 1994), has contributed a model monograph on what may prove to have been the last big Corps of Engineers construction project in the South. Luther J. Carter, *The Florida Experience: Land and Water Policy in a Growth State* (Baltimore, 1974), has woven politics, demography, and ecology together in a striking portrait of the farthest South, least southern state of Dixie. Together, these works provide an impressive introduction to the current range of environmental writing. The *Tulane Environmental Law Journal* offers an outlet for studies of one of the most legalistic and litigious movements in recent American history. Sources on laws and court decisions include Mary Robinson Sive, ed., *Environmental Legislation: A Sourcebook* (New York, 1976), and Oscar S. Gray, *Cases and Materials on Environmental Law* (Washington, D.C., 1977).

The view from ground level (or sea level) must be sought elsewhere; citations to a number of articles on forests, land, and wildlife are found in the notes to chapter 9. Recommended are the papers contained in the annual *Proceedings of the Southeastern Association of Fish and Wildlife Agencies*, which form a rich source on the lives and management of southern fauna and a useful corrective to the conventional rhetoric of decay, destruction, and loss. Articles of more than passing interest on recent environmental issues include: S.B. Cannon et al., "Epidemic Kepone Poisoning in Chemical Workers," *American Journal of Epidemiology* 107 (June 1978): 529-37; Gladwin Hill, "Ixtoc's Oil Has a Silver Lining," *Audubon* 81 (Nov. 1979): 150-59; Oliver A. Houck, "The Fish and the Dam," *Sierra* (May-June 1979): 15-17; and James Nathan Miller, "Trickery on the Tenn-Tom," *Reader's Digest* (Sept. 1978): 138-43. These articles cover a wide range in manner as well as subject matter: Cannon's is clinical, Miller's polemic, and Houck's a sardonic *vers libre* poem. The southern shoreline with which this book begins and ends forms the subject of a bewildering array of local studies, reflecting perhaps the "site-specific" nature of change there. Two widely separated parts of the southern coastline exhibit disparate but related problems in Vincent Bellis et al., *Estuarine Shoreline Erosion in the Albemarle-Pamlico Region of North Carolina* (Greenville, N.C., 1975), and R.D. Adams et al., *Shoreline Erosion in Coastal Louisiana: Inventory and Assessment* (Baton Rouge, 1978). Thomas J. Schoenbaum, *Islands, Capes, and Sounds: The North Carolina Coast* (Winston-Salem, 1982), writes ably of his topic, mingling history and ecology in an account of the state's program for managing its complex, historic coastline. Redding S. Sugg, ed., *The Horn Island Logs of Walter Anderson* (Memphis, 1973), with Sugg's introduction and Anderson's watercolors, enables the reader to see a part of the shore through the eyes of an artist.

Index

Abbot, Henry L., 99, 122
Aedes aegypti (mosquito), 4, 131-32
African diseases, 21, 37
African Americans, 187. *See also* blacks; slavery
Agricultural Adjustment Act, 164-65
agricultural experiment stations, 130
agricultural extension program, 109
agricultural reform, 58, 77-78
agriculture, 13-14, 16, 55, 58, 65, 67-68, 75-80, 104, 107-11, 141, 154-55
Agriculture, U.S. Department of, 130, 135, 156, 163-64, 180
Alabama, 2, 7, 11, 13, 72-73, 76, 79, 90, 94, 96, 98-99, 113-14, 120, 130, 134, 137, 139-40, 143, 152, 160, 162, 164, 177-78, 183
Alabama Industrial Development Board, 142
Alexandria, La., 160
Allegheny Mountains, 91
Allied Chemical Corp., 187
Altamaha River, 40
American Association for the Advancement of Science, 119
American Forestry Association, 119, 136
American Forestry Congress, 119, 135
American Ornithologists' Union (AOU), 116, 140
American Philosophical Society, 58
Anderson, Walter Inglis, 199-200
animal deviation, in malaria, 106
animals, 1, 12-13, 15, 17-18, 23, 26, 32, 45-50, 55-57, 69, 73, 94-95, 114-15, 178-81. *See also by kind*

anopheline mosquitoes, 4, 26, 34, 40, 48, 85, 106-7, 132, 154, 157, 170
Antebellum South, economy of, 79
AOU. *See* American Ornithologists' Union
Appalachian Mountains, 1, 65, 70, 72, 90, 120, 137
Appalachian National Park Association, 13, 16, 18
Archaic tradition, 13, 16, 18
Archdale, John, 42
Arkansas, 17, 70, 73, 84, 94, 98, 109, 113, 115, 117, 134, 139, 141, 164, 177, 186
Arkansas River, 73, 84, 97, 144, 185
Asheville, N.C., 136
Ashley River, 35, 40
Atchafalaya River, 92-93, 145, 149-50, 194
Atlanta, Ga., 86, 115, 158, 177, 186
Audubon, John James, 60, 99-101, 116
Audubon Society, 116, 138-40, 189
Austin, Moses, 67, 74
automobile industry, 186
Avery Island, La., 140
Ayllon, Lucas Vasquez de, 21

Bachman, John, 100
Baird, Spencer F., 115
Baker, Howard, 184
Baker, Willis M., 154
Baltimore, Md., 76, 86
Baltimore and Ohio Railroad, 96
Banister, John, 60
barrier islands, 4, 34, 191, 199
Bartram, John, 61

Bartram, William, 62-63
Baton Rouge, La., 85, 187
bauxite industry, 186
Beauregard, Pierre G.T., 97
beaver, 46-48, 179
Bennett, Hugh Hammond, 145, 159, 163-64
Berkeley, Sir William, 26, 51
Berkman, P.J., 108
Beverly, Robert, 50
Bienville, Jean Baptiste LeMoyne, Sieur de, 70
Big Bend National Park, 160
Big Cypress reservation, 162, 177, 192
Big Thicket, 176
Biloxi, Miss., 68
Biltmore estate, 119-20
Biltmore Forest School, 119
Biological Survey, Bureau of the, 116, 160
birds, 1, 4, 18-19, 45, 60, 92, 115-18, 138-41, 178-79, 189
Birmingham Ala., 114, 185
bison, 12-13, 18, 23, 46, 67, 94-95
black lung disease, 185
blacks, 21, 26-27, 34, 37-39, 87, 104-5, 111. *See also* slavery
Blackstone, William, 46
Blackwater River, 27
Bluegrass region, 2, 67, 79
boards of health: national, 104; state, 104-5, 134, 157
Bogalusa, La., 161
boll weevil (*Anthonomus grandis*), 110-11, 127-31, 139, 172-74
Boll Weevil Research Laboratory, 173
Bolzius, Johann Martin, 55
Bonnet Carré bend and spillway, 150-51
Boone and Crockett Club, 117, 119, 138
Boonesborough, Ky., 66
Borgne, Lake, 144
botany, 60-61
Brazos River, 74, 128
Breaux, John, 195
Broad Arrow policy, 53
Bromme, Traugott, 88
brown lung, 185
Buffalo Bayou, Tex., 142
Burroughs, John, 116
Byrd, William, II, 16, 27, 37, 48, 51, 57, 59-61
byssinosis, 186

Cable, George Washington, 87
Calcasieu River, 93, 109

Calhoun, John C., 96
Calvert, Cecilius, 33
Calvert, George, 33
canals, 96
cancer, as related to pollution, 188
Cape Fear, N.C., 3, 34-35, 40, 49
Cape Fear Arch, 3
Cape Fear River, 15, 21, 34, 53
Cape Hatteras, N.C., 35
Cape Henry, Va., 18, 34
Cape Lookout, N.C., 18
Carolina Sandhills National Wildlife Refuge, 162
Carter, Henry Rose, 132
Carter, Jimmy, 181, 184, 187
Carter, Landon, 58-59
Carter, Luther, J., 192
Cary, Austin, 137
Catawba Indians, 42
Catesby, mark, 42, 60
cattle, 23, 34-35, 49-50, 57, 69, 73, 76-77, 83, 85, 91, 106, 109-10, 113, 155, 162, 172, 180-81
CCC. *See* Civilian Conservation Corps
Charity Hospital (New Orleans), 88
Charles Town (Charleston), S.C., 35-37, 39-42, 51, 59, 61-62, 86, 100
Carlevoix, Pierre de, 16, 69
Carlotte, N.C., 138, 186
Charlotte Harbor, Fla., 140
Chateaubriand, François René de, 63
Chemin-a-Haut State Park, La., 160
Cherokee Indians, 41-52, 52, 84, 183-84
Chesapeake region, 3, 27, 31, 33, 66, 86, 95, 189, 191
Chevron Oil Corp., 189
Chicot State Park, La., 160
cholera, 86, 104
Civilian Conservation Corps (CCC), 151, 153, 159-63
Claiborne, William C.C., 70, 72
Clayton, John, 61
Clean Water Act, 194
Clearwater Harbor, Fla., 117
Cleveland, Grover, 119
climate, 2, 4, 62, 69-70, 85, 154
coal and coal industry, 5, 136, 188
Coleridge, William Taylor, 63
Commerce Clause of U.S. Constitution, 97, 122
Commerce Hills, 67
Communicable Disease Center, 170
Compagnie de l'Occident, 68

Congeree Indians, 41
conservation: of wildlife, in colonial South, 46-47; of land, in colonial South, 55-59; of forests, for U.S. Navy, 90; general lack of, in antebellum South, 94-95; private, attempted, 95; in Gilded Age, 103, 115-18; of forests, 118-20, 127; conservation movement, 135-36; of forests in 20th century, 136-38; of wildlife, 138-41; during 1920s, 142-43; under TVA, 153-54; of forests, in recent decades, 161; of wildlife, 161-62
Coolidge, Calvin, 143
corn, 5, 13-14, 16, 29, 32, 55, 58, 67, 75, 77, 130, 155
Corps of Engineers, U.S. Army, 121, 136, 144-45, 149-52, 158, 177, 185-86, 193-95
Corpus Christi, Tex., 189, 191
cost-benefit ratio, 151
cotton, 30, 67, 71-78, 103, 107-8, 110-11, 121, 123, 128-31, 154-55, 164-65, 169, 171-74
cotton gin, 71
Creek Indians, 52, 66, 72
Cross-Florida Canal, 192
Crown Zellerbach Co., 177
Cumberland River, 67, 115
Curtin, Philip D., 7
cutoffs on Mississippi River, 69, 97, 150
cypress, 92-93, 98, 113

Dale, Sir Thomas, 26
Darwin, Charles, 12, 43
DDT, 5, 158, 170, 172-73
deer, 5, 13-15, 23, 29, 35, 46-48, 52, 56-57, 67, 94, 138, 162, 179
Delaware, 15
dengue, 38
DeSoto, Hernando, 17, 20-21
disease, 4-5, 7, 19-22, 25-28, 34, 37, 38-39, 104-7, 128. See also epidemics; diseases by name
Dismal Swamp, 37, 48
District of Columbia, 117. See also Washington, D.C.
diversification of agriculture, 107-8, 110
Dow Chemical Co., 187
Drake, Daniel, 84, 100, 170
drug addiction, 104
Duffy, John, 26, 41
Duke University, 181
dysentery, 25, 85, 105-6

Eads, James B., 121-23
earthquakes, 3, 71
Ellett, Charles, 99
Elliott, Stephen, 100
Elliott, William, 94-95
endangered species, in colonial law, 56
Endangered Species Act, 180, 184-85
energy production, 152-53, 155, 183
English explorers, 18
Environmental Defense Fund, 184
environmental movement, 182
Environmental Protection Agency (EPA), 173, 181, 183
epidemics, 21-22, 25-26, 59, 87-88, 104-5
erosion: inland, 2, 32, 76, 79, 129, 152-53, 163-64; of shoreline, 3, 190-91
Evelyn, Robert, 26
Everglades, 4, 92, 98, 141, 191-93
evolution, varieties of, 6
Exxon Valdez, 190
extinctions, 11-12, 114-15, 178-81

Farm Credit Administration, 165
Farm Security Administration, 165
Faulkner, William, 197, 199
feather trade. See plume hunting
fencing, 54, 89
Ferguson, Harley B., 150
fertilizers, 14, 32, 58, 76-78, 103, 108, 152, 155
filariasis, 37, 39
Filson, John, 66
fire, 4, 14-15, 23, 35, 54-55, 94, 113, 176-77
fire ant, 5, 181
fish, 45, 92, 115-16, 153, 161, 178, 181, 183-85
Fish and Wildlife Service, U.S., 153
Fite, Gilbert C., 171
floods and flood control, 68-69, 95, 97-98, 108, 120-24, 143-44, 150-52
Florida, 1-2, 4, 7, 13, 18, 21, 35, 51, 53, 62-63, 72, 74, 90, 92, 94, 98, 112, 116-18, 127, 133, 139-42, 162, 177-80, 191-93, 198
Florida East Coast Railroad, 141
Florida Game and Freshwater Fish Commission, 192
food supply, 12-14, 23, 25, 78-79, 94, 106, 114, 156-57, 169-70
Ford, Gerald, 188
Forest Farmers Association, 176
forestry, scientific, 119-20, 137-38, 155, 160-61, 174-78

236 INDEX

Forestry Relations Department (TVA), 153
forests, 2, 5, 14-15, 29-31, 52-55, 89-91, 94-95, 111-14, 118-20, 174-78
Forest Service, U.S., 137, 145, 153, 159,
Fort Smith, Ark., 84
Frady, Marshall, 178
Frankfort, Ky., 113
Franklin, Benjamin, 62
freedmen, 111
fur trade. *See* skin trade

Galveston, Tex., 191
Garden, Alexander, 39, 59, 61-62
Garfield, James, 122
gas, natural, 188
Gentleman of Elvas, 21
Georgetown, Ky., 186
Georgia, 7, 13, 21, 36, 38, 51, 53, 62, 66, 72-75, 90-91, 94, 114-15, 129-30, 133, 137, 140, 152, 158, 176-77
Georgia, University of, 21
Georgia-Pacific Co., 177, 187
General Superfund, 188
"Germ of laziness." *See* hookworm disease
Gibbons v. *Ogden*, 96
Gibson, Randall L., 121-24
Gingaskin Indians, 27
glaciers, 2-3, 13
Goldberger, Joseph, 156-57
Gore, Albert, 182
Gorgas, William Crawford, 131
Grant, Ulysses S., 97
Great Smokies National Park, 137
Great Smoky Mountains, 6, 183
Great Valley, 2, 7, 66-67, 75, 77, 79
Greensboro, N.C., 138
Grinnell, George Bird, 117
guano, 77-78
Gulf of Mexico, 1, 3-4, 68

Hakluyt, Richard, 18
Hardtner, Henry E., 137
Harriot, Thomas, 18-20, 22
Harrodsburg, Ky., 66
Hatch Act, 130
Hayes, Rutherford B., 122
headright system, 30
Hearn, Lafcadio, 4, 196-97
Halper, Hinton Rowan, 195
Herty, Charles Holmes, 137
Hinns, John, 58
Hiwassee River, 183-84

hog, 50, 57, 69, 106
Holbrook, John E., 100
Holmes, Joseph A., 120
hookworm disease, 4, 37-38, 133-35, 166, 169
Hoover, Herbert, 144-45
Hopewell, Va., 187
Horn Island, Miss., 199
horse, 12-13
Houston, Tex., 2, 142
Howard Associations, 88
Humphreys, Andrew A., 99, 122
hunting, 5, 11-14, 46-50, 55-57, 73, 78, 94-95, 114-18, 138
hurricanes, 4, 71, 128, 141, 191

Ickes, Harold, 160, 162
immigration, 88
Indians, 11-21, 23, 31, 41-43, 46-47, 51-52, 69-70, 76, 84. *See also tribes by name*
industry, 103, 186-87. *See also industries by product*
insecticides, 110-11, 130-31, 172-73
insects, 4, 29, 68, 109-11, 127-31, 162, 172-74, 181. *See also Aedes aegypti; Anopheles;* boll weevil; fire ant
Interior, U.S. Department of the, 159-60, 162, 164, 178
internal improvements, 96-97
International Harvester Co., 165
International Paper Co., 177
intracoastal waterway, 193
iron industry, 29-30, 79, 111, 114
islands, barrier, 4, 34, 191, 193, 199

Jackson, Andrew, 96
Jamestown, Va., 4-5, 25, 27, 154
James River, 30
Janney, Israel, 58
Jefferson, Thomas, 58-59, 77, 100
Johnson, Lyndon, B., 182

Kentucky, 2, 13, 66-67, 75, 84-85, 94-99, 113, 115-16, 139, 152, 183, 188
Kepone, 186-87
Keys, Florida, 1
Kimberly-Clark Co., 177
Kirby Lumber Co., 135
Kisatchie National Forest, La., 160
Kissimmee River, 193
Knapp, Seaman, A., 109, 128
Kolomoki, Ga., 13
kudzu, 180

labor, in tobacco planting, 31
Lacey, John F., 118
Lake Charles, La., 113
Lamar, Lucius Q.C., 119, 122
landscape, in art, 6-7
Lane, Ralph, 19
Lawson, John, 33, 41, 47-48, 50, 61
League of American Sportsmen, 118
Lee, Robert E., 97
Leesburg, Va., 132
Leopold, Aldo, 200
Le Page du Pratz, Antoine Simon, 47, 58, 69-70
levees, 3, 5, 68-69, 95, 97-99, 108, 110, 121-24, 143-45, 194
"Levees-only," 99, 123-24
Lexington Ky., 100
life expectancy, in early Chesapeake, 28, 33-34
Life Science Products Co., 187
Lilienthal, David, 153, 155
Little, Henry, 42
Little Tennessee River 183-84
Llano Estacado region, 11
Long, Huey P., 142-43, 160
Longleaf pines, 90
lost colony, 22
Louisiana, 7, 68-71, 73-74, 88-90, 92, 94, 98, 108-9, 111, 113, 116, 121, 123, 128, 137, 139-41, 143-44, 149-52, 158, 160, 162, 171-72, 179, 187-88, 193-95
Louisiana Commission for the Conservation of Natural Resources, 137
Louisiana Purchase, 67
Louisiana Sugar Experiment Station, 108
Louisiana Sugar Planters Association, 108
Louisville, Ky., 67, 96, 106, 113
Louisville and Nashville Railroad, 184
Lufkin, Tex., 161
lumbering industry, 89-90, 93, 112-13, 136, 161
Lumber Manufacturers Association, 174
Luray, Va., 159

Macon, Ga., 162
Madison, James, 58
malaria, 4, 20, 26-28, 37-38, 40-41, 48, 68, 74, 83, 85-86, 104, 106-7, 109, 131-32, 134-35, 153-54, 157-59, 170-71
Malaria Control in War Areas, Office of, 158-59, 170
Marshall, John, 96

marshes, *See* wetlands
Maryland, 7-8, 33, 49, 56-57, 66, 72-73, 75-77, 162, 189, 191
Maryland, University of, 181
Matthiessen, Peter, 184
Maury, Matthew F., 100
McFarland, N.C., 119
McIlhenny, E.A., 140
medicinal plants, 19
medicine, 59, 99-100, 104
Memphis, Tenn., 97, 104-5, 144
Mencken, H.L., 197
Miami, Fla., 127, 141-42, 192
Miami, University of, 181
Michaux, André, 100
Mississippi (state), 2, 7, 71-73, 90, 96-98, 111, 113-14, 121, 123, 128, 130, 139-40, 161, 177, 179, 199
Mississippian tradition, 13, 16-17, 68, 70
Mississippi River and Tributaries Project, 149-52
Mississippi River and valley, 2-3, 5, 13, 17, 51, 65, 67-74, 84, 92, 95-97, 99, 104-5, 110, 113, 120-24, 143-45, 149-52, 185, 193-94, 198-99
Mississippi River Commission (MRC), 121-24, 143-45
Mississippi State University, 173
Missouri, 2, 13
Mobile, Ala., 86, 181
Mohr, Charles, 113
Monongahela River, 66
Monroe, La., 113
Montgomery, Ala., 178
Moravians, 35
Morgan, Arthur, 155
Morgan, Harcourt, 153, 155
Morgan City, La., 140
mortality: among white immigrants, 33, 69, 70, 87-89; among blacks, 38-39
mosquito. *See* anopheline mosquitoes; *Aedes aegypti*
Mount Landing, Miss., 144
MRC. *See* Mississippi River Commission
Muir, John, 118, 198
Muller, Paul, 158
Muscle Shoals, Ala., 152, 155-56

Nachitoches, La., 69
Nacogdoches, Tex., 93
Nader, Ralph, 175
Nashville, Tenn., 66, 113

238 INDEX

Nashville, Basin, 67, 72
Natchez, Miss., 66, 73, 128-29
Natchez Indians, 68, 70
National Board of Health, 104
National Cotton Council, 173
National Environmental Policy Act, 185, 194
national forests and parks, 120, 136-37, 159-60
National Game, Bird, and Fish Protection Association, 118
National Guard, 151
National Malaria Eradication Program, 170
National Oceanic and Atmospheric Administration, 190
national parks, 136-37, 160
National Park Service, 153, 191
National Wildlife Federation, 192
naval-stores industry, 34-35, 40, 53, 90, 113, 137
navigation, of rivers, 69, 95-97
Negroes. *See* blacks
New Bern, N.C., 61
New Madrid, Mo., 3, 71
New Orleans, La., 59, 68, 70, 75, 86-89, 95, 97, 104-5, 115, 131, 143-45, 150, 187, 191
Nicholson, Francis, 60
Nixon, Richard, 182
Norfolk, Va., 86, 115
Norris, George, 152
North American Land and Timber Co., 109
North Carolina, 13, 15, 19, 22, 33-34, 37, 48, 53-54, 57, 59, 61-62, 72-73, 75, 90, 93, 108, 120, 136-38, 143, 152, 160, 162-64, 171-72, 176-77, 186
North Carolina, University of, 119
North Carolina Department of Conservation and Development, 142
Nott, Josiah C., 86, 88
Nottoway Indians, 27, 51
Nuttall, Thomas, 84, 100

Oak Ridge, Tenn., 156
Ocmulgee River, 66
Oconee River, 66
Oglethorpe, James Edward, 36
Ohio River and valley, 2, 12, 13, 66-67, 73, 84, 96
Oil Conservation Board, 143
oil industry, 136, 141-42, 187-90
Okeechobee, Lake, 141, 192
Okefenokee Swamp, 192
Oklahoma, 2, 72, 134, 141, 143, 177

Orlando, Fla., 140
ornithology, 60. *See also* birds
Ouachita River, 113, 172
Ozark Mountains, 72, 113

Paleo-Indians, 11-13, 16
parks: national, 136-37, 160; state, 159-60
passenger pigeon, 45-47, 95, 114-15
Patuxent Wildlife Research Center, 189
peanuts, 171-72
Peal River, 7, 92
Pearson, Thomas Gilbert, 138-39
pellagra, 156-57, 169
Pemlico Indians, 41-42
pesticides. *See* insecticides
petrochemical industry. *See* oil industry
Petroleos Mexicanos, 189
Petulla, Joseph, 200
Phillips, Ulrich B., 7
Piedmont region, 6, 32, 66, 72, 75-77, 83-84, 91, 129, 163, 176-77
Pinchot, Gifford, 119-20, 135-36, 164
pines, 2, 15, 90, 91, 112-13, 137, 160-61, 174-75
place, sense of, 6
plantations, 30, 107, 110
plants, 19, 23
plasmodium, 40, 85
Pleistocene epoch, 3, 11-12, 20, 114, 200
plume hunting, 117-18, 138
podzols, 2
pollution, 5, 26, 172, 174, 178, 181, 186-90
Pontchartrain, Lake, 2, 145, 150
population, 14, 19-20, 26-27, 33, 39, 46, 65, 69, 73, 78
Potomac River, 2
poultry, 172
Poverty Point, La., 13
Powhatan Indians, 27
predators, 49-50, 94
Public Health Service, U.S. (USPHS), 132-33, 156, 158, 170
Public Works Administration (PWA), 151
pulp and paper industry, 160-61, 174-78

quarantines, 87, 104-5
quinine, 85

Rafinesque, Constantine, 100
railroads, 95-96, 113. *See also by name of line*
rainfall, 2, 79
Raleigh, Sir Walter, 19, 34
Raleigh, N.C., 115, 138

Ransdell, Joseph E., 143
Ravenel, Henry W., 100
Rawlings, Marjorie Kinnan, 198
Red Cross, 151
Red River, 73-75, 92-93, 97, 121, 123, 172
Reed Walter, 106-7, 131
resources, 18, 28, 79
respiratory diseases, 38-39
rice, 23, 35, 38, 40-41, 68, 74-75, 87, 108-9, 171
Riceboro, Ga., 175
Rio Grande, 134, 173, 190
River basin planning, comprehensive, 143-45, 152-56
rivers, development of, 95-98. *See also* internal improvements; levees; Mississippi River; navigation; river basin planning
rivers and harbors acts, 14, 151
Roanoke Island, N.C., 18
Rockefeller Foundation, 132, 140
Rockefeller Sanitary Commission, 133-34
Roland, Charles P., 197
Rolph, John, 30
Roosevelt, Franklin D., 152, 162
Roosevelt, Theodore, 117, 120, 135-36, 139-40
Rubin, Julius, 77
Ruffin, Edmund, 78
Russell Sage Foundation, 140

Saint Augustine, Fla., 21
Saint Francis River, 68, 97
Saint Johns River, 62
Saint Louis, Mo., 13, 106
Saint Louis Merchants Exchange, 123
Saint Mary's, Md., 33
Salem, N.C., 35
Saline River, 113
Savannah, Ga., 86-87, 161
Savannah River, 21, 36, 175
Savitt, Todd L., 85
Schenck, Carl A., 119
Schurz, Carl, 119
science, 6, 59-63, 98-101
Scott, W.E.D., 117
Scott Paper Co., 177
scurvy, 69
Sea Islands, 3, 35
sea levels, 3, 193
Sears, William H., 7
Seminole Indians, 74, 85, 162
Sewee Indians, 42
sharcropping, 107

Shell Oil Co., 189
Shenandoah valley, 66
shores, 34, 190-95
Shreve, Henry M., 97
sickle-cell anemia, 39
silk culture, 29, 36
Singleton, John, 58
skin trade, 35-36, 49-52, 67, 94
slavery, 34-39, 65, 68-69, 74-75, 78, 83, 85, 100
smallpox, 20, 26-27, 34, 41-43, 46, 59, 62, 70
Smith, John, 27, 29
Smithsonian Institution, 115, 121
Smyrna, Tenn., 186
snail darter, 184-85
Soil Conservation Service, U.S., 145, 159, 163-64, 177
Soil Erosion Service, U.S., 164
soils, 2-3, 30-31, 34, 58, 68, 73-79, 154-55
South Carolina, 11, 35-36, 39-40, 42, 49-51, 57, 61-62, 66, 73, 75-76, 94, 109, 129-30, 133, 139, 161, 177
South Carolina Natural Resources Commission, 142
Southern Forestry Congress, 137
Southern Homestead Act, 111-12, 120
Southern Pacific Railroad, 113
Southern Pulpwood Conservation Association, 161
Southern Railroad, 136
Southern Tenant Farmers' Union, 165
South River, Md., 32
South-Western Convention, 97
soybeans, 169, 171, 177
Spanish-American War, 106
Sparks, William A.J., 119
Sport Fishing Institute, 181
Spring Hill, Tenn., 186
Starkville, Miss., 162
Stiles, Charles Wardell, 133
Strachey, William, 47
stripmining, 188
sugar, 23, 37, 68, 108, 121, 123
surveys, 99, 120
Swamp Lands Acts, 92-93, 98

Taft, William Howard, 136
Tamiami Trail, 192
Tarpon Springs, Fla., 117
Taylor, John, 58, 77
Tchefuncte State Park, La., 160
tenancy, 107, 165

Tennessee, 7, 13, 66-67, 72-73, 85, 94, 137, 140, 152, 160, 182-86
Tennessee, University of, 184
Tennessee River, 67, 152-56, 183, 185
Tennessee-Tombigbee Waterway, 185-86
Tennessee Valley Authority (TVA), 152-56, 182-85
Tensas River, 123
teosinte, 13
Texas, 2, 4, 7, 11, 13, 18, 23, 73-75, 77, 93-94, 109-10, 113, 118, 128, 134, 141-43, 160-61, 165, 177, 188-91
Texas fever, 109-10
Thomas, Norman, 165
Tidewater region, 30, 32, 58, 76, 89
tobacco, 19, 29-33, 35, 58, 66-68, 164
Tonti, Henri de, 69
tornadoes, 3, 4
Transappalachia, 66
Transfer Act, 135
Transylvania University, 100
tree farming, 171, 174-78
Trimble, Stanley, W., 76
tuberculosis, 170
Tugwell, Rexford, 155
Tulane University, 121
tung oil, 171
turkey, 46-47, 95, 138, 179
turpentine, 90
Tuscarora Indians, 61
Twain, Mark 198-99
typhoid, 25, 68, 84, 106, 154

Union Bag and Paper Co., 176
Union Camp Co., 177
USDA. *See* Agriculture, U.S. Department of
USPHS. *See* Public Health Service, U.S.

Vaca, Cabeza de, 18
Vanderbilt University, 134
Vicksburg, Miss., 13, 145, 150
Virginia, 13, 15, 18, 25-34, 48-51, 53, 56-57, 59-60, 66, 72-73, 75-76, 85, 89, 95, 106, 108, 117, 132, 137, 139, 143, 152, 160, 187, 191, 197
Virginia Co., 28, 30
Virginia Society for the Promotion of Useful Knowledge, 58

Wallace, George, 58, 175
Washington, D.C., 86
Waterways Experiment Station, 145
West Virginia, 91, 116-17, 139, 188
wetlands, 2, 3, 48, 91-95, 98, 149-50. *See also* Atchafalaya River; Dismal Swamp; Everglades; Okefenokee Swamp; Swamp Lands Acts
Weyerhaeuser Lumber Co., 135, 174, 177
wheat, 23, 32, 35, 58
White, John, 18
White House Conference of Governors (1908), 135
wildlife. *See* animals; birds; extinctions; fish
wildlife refuges, 161-62
Wilson, Alexander, 100
Wilson Dam, 152
Winston, N.C., 138
wolves, 49-50, 56, 94
Woodland tradition, 13, 16
Woodman, Harold D., 79
wood products industry, 29, 174-78. *See also* forestry, scientific; forests; naval-stores industry; pulp and paper industry
Works Progress Administration (WPA), 157-58

yaws, 4, 37, 39, 69, 83
Yazoo River, 97, 110, 121, 123, 128
yellow fever, 4, 34, 37-38, 70, 85-87, 103-5, 127, 131
Yellow Fever Commission, 131

Zea mays. *See* corn